Coding for Kids Ages 8-12

Table of Contents

Disclaimer

Introduction

It was a cool, windy evening in Chicago. I was enjoying the playoffs on television. Suddenly, I heard a crackle on the TV, and the picture started getting fuzzy. A couple of minutes later, the whole screen went green. No picture, no game. It was frustrating to have the TV go out in the middle of the game.

Hearing my groan of annoyance and frustration, my wife, Candy asked me what happened. 'Well. The TV screen just went blank. Right in the middle of the game'. My 9-year-old son, Steve, who just walked in the room, said 'That's an easy fix'. Steve switched the television off and made a few adjustments to the satellite box. And two minutes later, he switched the television on, and it was working. Like magic!!!

Next, we have the case of Abby. She was a tennis player in Madison high school. She had been struggling for months since she recovered from an elbow injury. She'd lost 3 tournaments this year which she had won the previous year. She was upset, but she realized that the loss happened for a reason. It wasn't bad luck or lack of skill. It was due to certain circumstances. She decided that she needed to change things quickly. She spent more time analysing replays of her game and how it had changed in the last six months. She figured out that her grip on the racket had changed since the injury and her forehand was badly affected. She changed her grip at practice and instantly felt she was playing a lot better. At the next tournament, she won easily. Abby was back on track.

Now, let's look at the final example of Matthew. Matthew was a 12-year-old in a small town in rural Wyoming. He had interest in science and technology, but there wasn't much opportunity to explore this in his community. One day, someone donated a dozen computers in his school, and he was one of the lucky ones to be able to take it home. The computer was pre-installed with C++. He was intimidated with the idea of coding. However, with the help of a few books in the library, he got started. He did some basic projects and eventually became the coding whiz of the school. His classmates loved his games and animations and asked him to teach them how to code. Matthew eventually went to college for computer science, and now works for a very high salary as a programmer in Silicon Valley.

All the above situations are hypothetical ones, where the innate ability of Matthew, Abby and Steve led to wonderful outcomes. They had the grit and determination to deal with adversity and learn quickly. However, there are other skills that each of them have shown that can be learnt with programming.

Steve had been learning programming for the last couple of years, and his problem-solving skills were advanced. He was able to break things apart, fix them and repair items like televisions and computers easily.

Abby had also learnt coding recently. She was really good at spotting mistakes in the code and fixing them. This had translated into real life in different situations, with an ability to spot details that others struggled with.

Matthew's life had changed due to programming. That was his only outlet for his skills and creativity in a community that didn't have many opportunities. He now has his dream career and hopes to inspire many more in his situation to do the same.

These above three examples show how coding can completely change one's mindset, problem solving abilities, creativity; and greatly improve one's quality of life.

What is Programming?

Programming is the creation of a systematic set of instructions that a computer can understand and execute. These instructions are called code, and they are written in a language that's known as a programming language. There are many extraordinary programming languages, each with their own personal syntax and capabilities.

Benefits of Programming

Programming is like a paintbrush for the digital world. It lets you to create and form thoughts into something tangible, something that can be shared and experienced by using others.

With programming, you have the strength to build and layout digital landscapes, from the tiniest of small print to the grandest of software designs. You can create interactive experiences that captivate audiences. You can automate tedious tasks, making them extra efficient and freeing up time for more important pursuits.

Programming is a great tool for problem-solving. It allows you to take complex issues and break them down into manageable pieces. It's like a jigsaw puzzle, the place you coalesce blocks collectively to build something greater. With programming, you can flip data into knowledge, and that knowledge into action. You can communicate with machines and make them do your bidding. It's like a secret language that only a select few are privy to, and with that knowledge, you can liberate hidden tasks.

Programming is a medium through which you can express your creativity, fix problems, and change the world. It's a tool that empowers you to carry your ideas to reality and make a real impact on the world.

It is an in-demand skill, with job possibilities in a diverse range of industries. From bookkeeping to cybersecurity, programming is a versatile tool that can lead to a beneficial career. Programmers are well-paid, and the demand for expert programmers is growing. According to the Bureau of Labor Statistics, the median salary for computer programmer in the United States is $88,240. It is not uncommon for people with 5 years of work experience to have a salary of $150000 or more.

Programming can be used to automate repetitive tasks, making them more efficient and less time-consuming. This can be utilized in a variety of industries, from manufacturing to finance.

It is a field that is constantly evolving, with new technologies and languages emerging all the time. Learning to code is a journey, and there is always something new to learn, which can be interesting and motivating.

Programming is also a collaborative effort and learning to program can help you learn how to work in a team. Understanding how to speak and work with others is a necessary skill, and programming can help you enhance that skill.

Programming abilities can also be used to start your very own business. Whether you choose to create an app, internet site or software, programming is an indispensable tool for constructing and launching a successful start-up. The world is lit with startups that have been launched by programming and grown into billion dollar businesses, like Facebook, Google, Amazon etc.

Programming empowers you to take control of technology and use it to obtain your goals. Instead of being a passive consumer of technology, you can become an energetic creator and use technology to make a tremendous have an impact on in your lifestyles and the lives of others.

Why Choose C++

Imagine you are a sculptor, with a block of stone in front of you. C++ is like the chisel in your hand, a precision device that approves you to carve and form your masterpiece with precision and control.

With C++, you have the energy to shape and form your code with great efficiency, making it run quicker and smoother. It's a low-level language, which has better communication with the machine. It allows you to fine-tune your program's performance, like a sculptor cautiously chiselling away the excess marble.

Think of other programming languages as paint brushes, they are amazing for creating colorful, expressive pieces; but they may additionally lack the precision and power of a chisel. C++ is like a Swiss Army knife of programming languages, it has equipment for unique purposes, like object-oriented and procedural programming.

C++ additionally has a significant and energetic community of developers, who continuously enhance and update the language, making it extra strong and versatile. This neighbourhood is like a crew of specialist sculptors, who are constantly inclined to share their knowledge and strategies with you.

C++ is a powerful, versatile programming language that is extensively used in enterprise and academia. It provides a solid foundation in programming standards and can be used to create a wide variety of software, including working systems, games, and simulations. Additionally, C++ is a compiled language, which potential that it can be optimized for performance and is commonly faster than interpreted languages. A compiled languages converts the entire code into machine language; and the computer runs the machine language. This is much faster than an interpreted language, which converts each line into machine language at run time.

C++ supports each object-oriented and procedural programming paradigm, making it flexible and versatile.

C++ is extensively used in industry and academia and is supported with the aid of most important running systems, giving it a high level of portability.

It's well worth noting that C++ is not always the best programming language for every application, and different languages may be better proper for certain tasks. However, C++ is a powerful and versatile language that can be used to create a broad range of software, and its strengths make it a popular choice among developers.

C++ Installation

Code Blocks is a software that we will use to execute C++ code and view the output. It is a beginner user-friendly IDE that is great to get started with it.

Steps

1. Browse to http://www.codeblocks.org/downloads/binaries/

Download the version that's best suited for your operating system by clicking on one of the links below:

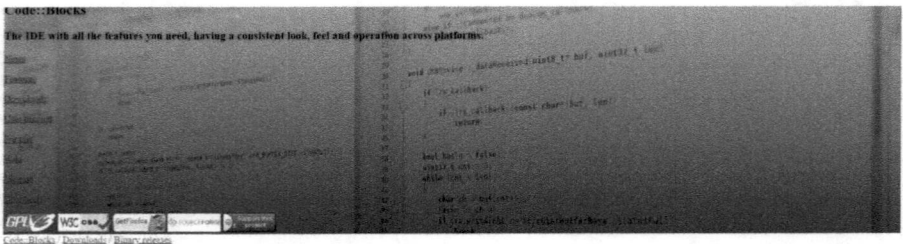

2. Install the program.

3. Once installed, open the Code::Blocks software below from the Start Menu.

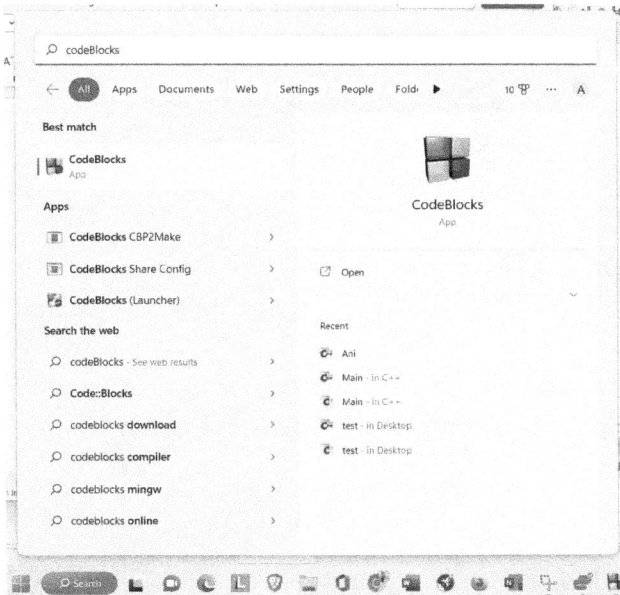

4. Open a new file as shown below.

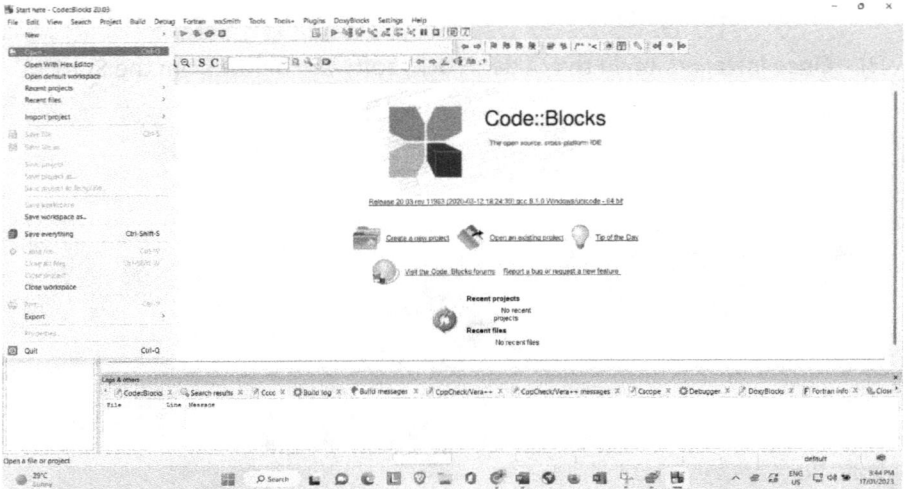

5. We're now going to test out some basic code below. Copy the code below highlighted in yellow and paste it in the file.

```
#include <iostream>

int main()

{

    std::cout << "Hello, World!" << std::endl;

    return 0;

}
```

It should look like this below:

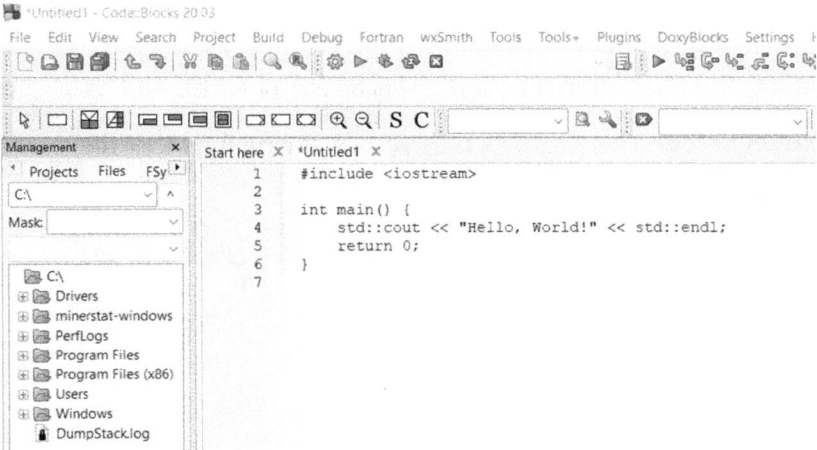

6. Now, click **"File > Save File As"**

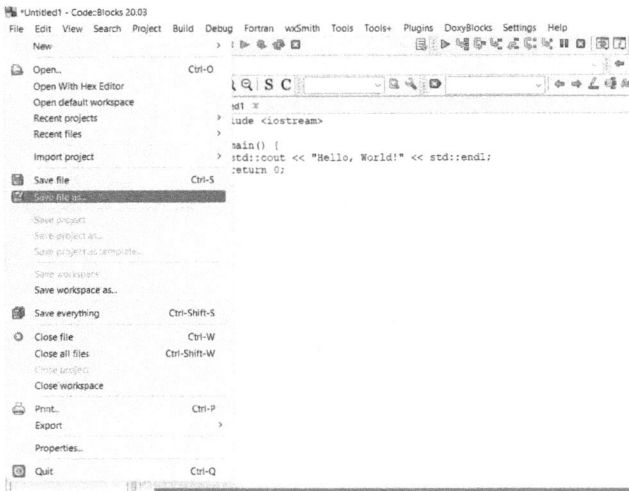

7. Name, the file and make sure it has a ".cpp" at the end as shown below. This ensures that you're saving it as C++ file (instead of a C file).

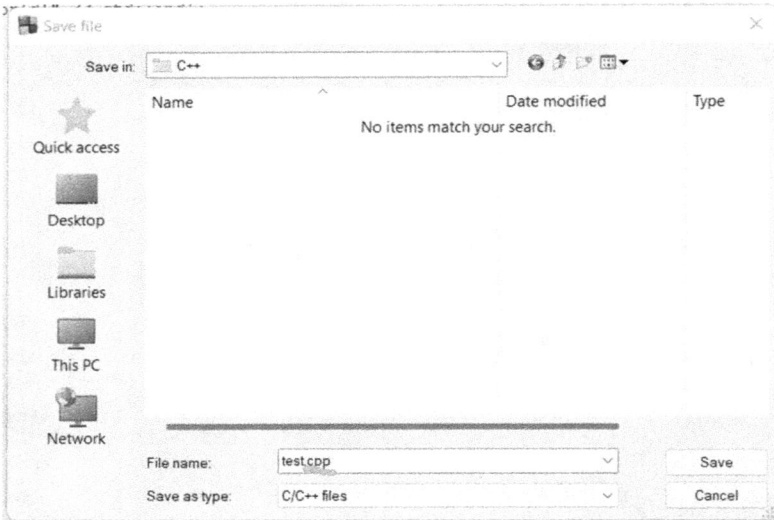

8. Now, build the code as shown below:

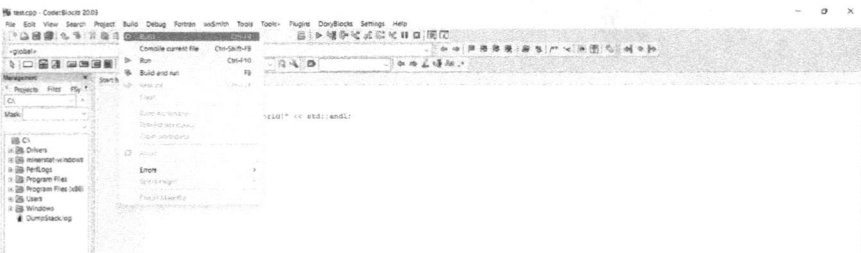

9. The log at the bottom shows whether build was successful. In this case, 0 errors with a successful build, as shown in picture below.

10. Once build is complete, run the code as shown below.

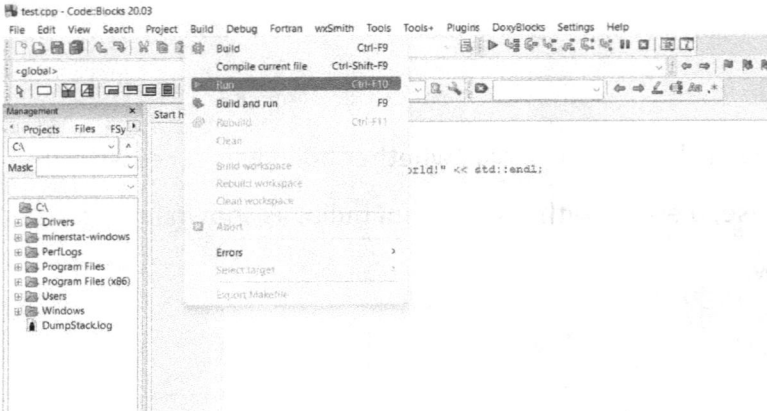

11. An output window should pop up with the message "Hello World" as shown below.

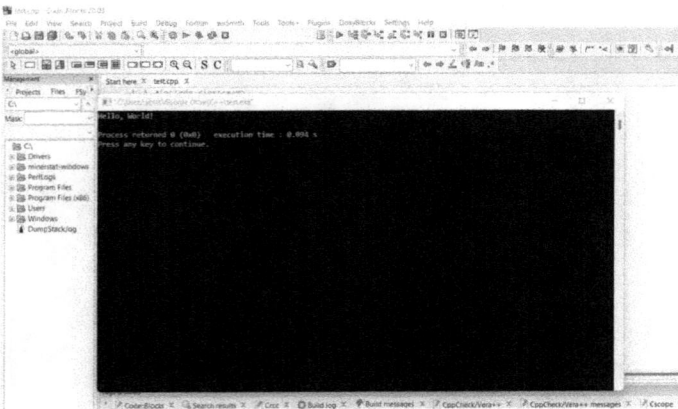

12. Press any key to exit window.

Simple C++ Programs to Start

In this chapter, we're going to go over a couple of basic programs to get you started. And explain every single line of the code.

This will help you understand the purpose and objective of the code.

Program 1: "Hello World"

The first program is a "Hello World". It's the first program that most programmers write. It helps beginner programmers get started with the syntax and should help you execute your first program in C++. If the code seems familiar, it's the same code that we went over in the "C++ Installation" chapter. Now, you'll be able to understand what the code means.

```cpp
#include <iostream>

int main()

{

    std::cout << "Hello, World!" << std::endl;

    return 0;
```

}

```
1    #include <iostream>
2
3    int main() {
4        std::cout << "Hello, World!" << std::endl;
5        return 0;
6    }
7
```

The purpose of the program above is to print the words "Hello World" on the screen.

The first line **"#include <iostream>"** imports a library called iostream into the program. It basically means that you can now use all the commands within iostream. **Iostream** contains two popular commands that this program will use, called **cout** and **endl**.

Cout allows you to print an output on the screen; and **endl** allows you to move to the next line of the output.

In Line 3, we have **"int main()"** which is used to start the actual program. The program executes every line after **int main()** which are included between the curly braces that look like "{}". These 2 braces indicate the start and end of the program.

In line 4, **"std::cout << "Hello, World!" << std::endl;"** uses the "cout" object from the iostream library to output the string "Hello, World!" to the console. The "<<" operator is used to insert the string into the output stream. The "**std::endl**" is used to insert a newline character after the string, so the next output will appear on a new line.

In line 5, "**return 0**" indicates that the program is complete by returning 0 to the operating system.

When we run the program, we get the message "**Hello World**", as shown below.

Output:

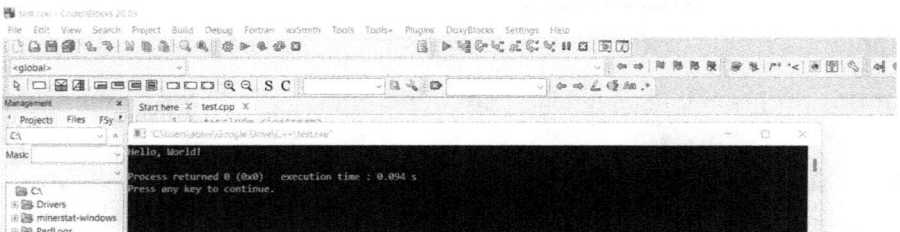

std::cout vs cout: Which One Do I Use?

Throughout the rest of the book, you will notice **std::cout** being used to print characters on the screen, instead of **cout**.

This is because **cout** is a member of the **std namespace**. If we don't use **std::cout** then there is a possibility of a naming conflict if other namespaces are used.

This is the same reason that we use **std::endl** instead of **endl**; and **std::string** instead of **string** in the rest of the book.

Program 2:

Program 2 is very similar to Program 1. Instead of printing out "Hello World", it prints out "Welcome to Coding in C++"


```
#include <iostream>


int main()
{
    std::cout << "Welcome to coding in C++" << std::endl;

    return 0;
}
```

here ✕   test2.cpp ✕

```cpp
1    #include <iostream>
2
3    int main()
4    {
5        std::cout << "Welcome to coding in C++" << std::endl;
6        return 0;
7    }
8
```

The only difference is in line 5. The text between << and << is

"Welcome to Coding in C++".

So, the code prints out "Welcome to Coding in C++"

Output:

```
"C:\Users\abhis\Google Drive\C++\test2.exe"
Welcome to coding in C++

Process returned 0 (0x0)   execution time : 0.012 s
Press any key to continue.
```

Problems to Practice

1. Write a program that prints your name on the screen.

2. Write a program that prints your name and age on the screen.

A Short message from the Author:

Hey, are you enjoying the book? I'd love to hear your thoughts!

Many readers do not know how hard reviews are to come by, and how much they help an author.

I would be incredibly thankful if you could take just 60 seconds to write a brief review on Amazon, even if it's just a few sentences!

Click this link and leave a review…

Thank you for taking the time to share your thoughts!

Your review will genuinely make a difference for me and help gain exposure for my work.

Variables

A variable is the most fundamental concept in programming.

It is basically a place to store a value. Think of it as a box where you can store something. In programming, you can store either a number, a string or a character.

A variable that can store a number is called an integer variable.

A variable that can store a string is called a string variable.

The value of a variable can change within a program. Variable in C++ are used to store and manipulate data.

In C++, we declare variables using the **var** command in iostream library.

Here's an example where we create a integer variable **num** and store the number 22 in it.

```cpp
#include <iostream>

int main() {
    int num = 22;
    std::cout << "The value of variable num is: " << num << std::endl;
    return 0;
}
```

In line 4, **"int num=22"** creates an integer variable **num** and assigns it a value of **22**.

In line 5, we print the value of num. See output below.

```
■ "C:\Users\abhis\Google Drive\C++\var.exe"
The value of variable num is: 22

Process returned 0 (0x0)    execution time : 0.169 s
Press any key to continue.
```

Now, let's add a couple of lines to above program to change the value of the variable.

```
1    #include <iostream>
2
3    int main() {
4        int num = 22;
5        std::cout << "The value of variable num is: " << num << std::endl;
6        num = 55;
7        std::cout << "The new value of num is: " << num << std::endl;
8        return 0;
9    }
10
```

```
■ "C:\Users\abhis\Google Drive\C++\var.exe"
The value of variable num is: 22
The new value of num is: 55

Process returned 0 (0x0)    execution time : 0.053 s
Press any key to continue.
```

We add lines 6 and 7. In line 6, "num=55" changes the value of num to 55, and the new value of num now is 55. In line 6, we print the new value of num. See output below.

User Inputted Variable

The command **"cin"** from iostream library is used to accept an input from the user. It is shown below:

cin >> x

The above line accepts x as an input from the user.

The input can be used in a C++ program.

In the example below, we'll prompt you to pick a number, and print the number on the screen.

```
1    #include <iostream>
2
3    int main() {
4        int yournumber;
5        std::cout << "Enter any number: ";
6        std::cin >> yournumber;
7        std::cout << "The number that you entered is: " << yournumber << std::endl;
8        return 0;
9    }
10
```

In line 5, we print the words **"Enter any number"** on the screen.

In line 6, you type in a number, and it stores it in the variable **"yournumber"**.

In the line 7, the number is printed on the screen.

See output below.

Output:

```
■ "C:\Users\abhis\Google Drive\C++\userinputvar.exe"
Enter any number:  78
The number that you entered is:  78

Process returned 0 (0x0)     execution time : 4.635 s
Press any key to continue.
```

User Input with 2 variables

In the next example, we'll accept two variables, one for a name and another for age, and print out both. **name** is a string variable and **age** is an integer variable.

```
1    #include <iostream>
2    #include <string>
3
4    int main() {
5        std::string id;
6        int age;
7
8        std::cout << "What's your name: ";
9        std::cin >> id;
10
11       std::cout << "How old are you?";
12       std::cin >> age;
13
14       std::cout << "You are " << id << " and your age is " << age << " years." << std::endl;
15       return 0;
16   }
17
```

36

In line 2 of above code, **"#include <string>"** allows you to use all string commands, including creation of string commands.

In line 5, we create a string variable called **id** to hold your name.

In line 7 and 8, you input your name and store it in the **id** variable.

In line 8 and 9, you store your age in the **age** variable.

Finally in line 14, we print out the name and age you entered; as shown in output below.

```
"C:\Users\abhis\Google Drive\C++\nameage.exe"

What's your name: Steven
How old are you?12
You are Steven and your age is 12 years.

Process returned 0 (0x0)   execution time : 10.363 s
Press any key to continue.
```

Problems to Practice

1. Write a program that creates two variables. The first variable holds your favorite color, and the second variable holds your name. Print out both your name and color in the same line.

2. Write a program that creates a variable that describes where you are going on your next vacation. Print out the variable on the screen.

Boolean Variables

A Boolean is a variable type that checks it a statement is true or false.

The Boolean contains either the value 'true' or 'false'.

It is declared as:

bool a = 'true'; or

bool a = 'false';

A 'true' value in C++ is output as 1 on the screen, while a 'false' value is output as 0.

It can also be assigned to a value through a statement:

bool a,b;

a = (5<7);

b = (9<0);

a is 'true'. When it is output on the screen, **a** returns a value of **1**, which stands for 'true' in C++.

b is false. When it is output on the screen, **b** returns a value of **0**, which stands for 'true' in C++.

Now, let's look at an example that defines this more clearly.

Boolean Example

In the code below, two integer variables **num1** and **num2** are assigned in lines 7-8. Two string variables are assigned in lines 9-10.

3 boolean variables a, b and c are declared in line 6.

Boolean variable **a** in line 11 checks if num1 is equal to num2. In line 14, it returns 0 if false and 1 if true.

Boolean variable **b** in line 12 checks if num1 is less than num2. In line 15, it returns 0 if false and 1 if true.

Boolean variable **c** in line 13 checks if str1 is equal to num2. In line 16, it returns 0 if false and 1 if true.

```
1    #include <iostream>
2
3    int main() {
4      int num1, num2;
5      std::string str1, str2;
6      bool a, b, c, d;
7      num1 = 7;
8      num2 = 10;
9      str1 = "Fear";
10     str2 = "Yell";
11     a = num1==num2;
12     b = num1<num2;
13     c= str1==str2;
14     std::cout << "Is num1 equal to num2? (1 for true and 0 for false) = " <<a << std::endl;
15     std::cout << "Is num1 less than num2? (1 for true and 0 for false) = " << b << std::endl;
16     std::cout << "Is str1 equal to str2? (1 for true and 0 for false) = " <<c << std::endl;
17     return 0;
18   }
19
```

Output

In the output below, **a** is false and returns 0 as **num1** is not equal to **num2**.

b is true and returns 1 as **num1** is less than **num2**.

c is false and returns 0 as **str1** is not equal to **str2**.

41

```
"C:\Users\abhis\Google Drive   ×    +  ∨

Is num1 equal to num2? (1 for true and 0 for false) = 0
Is num1 less than num2? (1 for true and 0 for false) = 1
Is str1 equal to str2? (1 for true and 0 for false) = 0

Process returned 0 (0x0)    execution time : 0.043 s
Press any key to continue.
```

Problems to Practice

1. Write a program takes utilizes a Boolean variable to check if a student's grade is pass or fail. The pass mark is 60 out of a 100.

String Variables

A string variable is a variable types that holds only characters or a sequence of characters. It is part of the <string> C++ standard library. The <string> library is used to create and manipulate string variables; and it's accessed using the line below:

#include <string>

The variable is assigned as below:

string a = "xyz";

The above line assigns the value "xyz" to string a.

You can change the string to a value of "bdc" using line below..

a="bdc";

Basic String Example

In our first example, the user inputs his name and it's stored in a string variable called name and the string variable is printed on the screen.

```
1    #include <iostream>
2    #include <string>
3
4    int main() {
5        std::string name;
6        std::cout << "What's your name? ";
7        std::cin >> name;
8        std::cout << "Hi, " << name << "!" << std::endl;
9        return 0;
10   }
11
```

A string variable called **name** is declared on line 5.

In lines 6-7, the user enters his **name** into the string variable.

In line 8, the string variable is printed on the screen.

Output:

"C:\Users\abhis\Google Drive\C++\string_basic.exe"

What's your name? Tim
Hi, Tim!

Process returned 0 (0x0)   execution time : 1.987 s
Press any key to continue.

In example above, the user enters his name as Tim.

Tim is stored in variable **name** and printed on the screen.

Joining Strings

Strings in C++ can be joined together by using the '+' operator.

In the below example, the user inputs two strings and combines them into a single larger string variable. The combined string is printed on the screen.

```
1   #include <iostream>
2   #include <string>
3
4   int main() {
5       std::string string1;
6       std::string string2;
7
8       std::cout << "Enter the first string: ";
9       std::cin >> string1;
10
11      std::cout << "Enter the second string: ";
12      std::cin >> string2;
13
14      std::string combstring = string1 + string2;
15      std::cout << "The joined string is: " << combstring << std::endl
16
17      return 0;
18  }
19
```

Two string variables, **string1** and **string2** are created in lines 5 and 6.

In Lines 8-9, the user inputs the 1st string and it's stored in string variable **string1**.

In Lines 11-12, the user inputs the 2nd string and it's stored in string variable **string2**.

In Line 13, the two strings are joined together and combined in string variable **combstring**. **combstring** is printed to the screen in Line 14.

Output:

```
"C:\Users\abhis\Google Drive\C++\Double_Strings.exe"
Enter the first string: abc
Enter the second string: xyz
The joined string is: abcxyz

Process returned 0 (0x0)    execution time : 4.223 s
Press any key to continue.
```

The user inputs the first string as **'abc'**; and the 2nd string as **'xyz'**.

The combined string is **'abcxyz'**, which is stored in variable **combstring** and printed on the screen.

Length of string

In C++, the length of a string is calculated using the getline and length functions. In the below code, the user enters a string and the program outputs the length of the string.

```
1    #include <iostream>
2    #include <string>
3
4    int main() {
5        std::string string1;
6
7        std::cout << "Enter the string: ";
8        std::getline(std::cin, string1);
9
10
11       std::cout << "The length of the string is: " << string1.length() << std::endl;
12
13
14
15       return 0;
16   }
17
```

In the program in line 8, the user enters the string using getline.

The reason we use getline instead of a normal cin prompt is that getline takes into account all spaces that are in the string. If we just use cin, the string stops at the first space, and we get an incorrect length.

In line 11, the string is printed on the screen as shown in output below, using **string1.length()**.

Output:

In the output below, the user enters the string **"I won the race"** and it is stored in string1. The **getline** function gets the entire length of the string and the **string1.length()** prints the length as **14**.

If we used **cin** instead of getline, then we get a length of 1 as the first space starts after 1 letter.

```
Enter the string: I won the race
The length of the string is: 14

Process returned 0 (0x0)    execution time : 3.240 s
Press any key to continue.
```

Substrings

In C++, a substring is a portion of a larger string. The location of the substring is found using the find() function. In below code, the program accepts two strings as input. It then looks to find the location of the smaller string in the larger one. The location is called the index of the string, which is the position within the string.

```
1    #include <iostream>
2    #include <string>
3
4    int main() {
5        std::string string1, substring;
6        std::cout << "Enter the main string: ";
7        std::getline(std::cin, string1);
8        std::cout << "Enter the substring to be found: ";
9        std::getline(std::cin, substring);
10
11       int index = string1.find(substring);
12       if (index >=0) {
13           std::cout << "The substring \"" << substring << "\" was found at index " << index << " in the main string." << std::endl;
14       } else {
15           std::cout << "The substring \"" << substring << "\" was not found in the main string." << std::endl;
16       }
17
18       return 0;
19   }
20
```

In the example, the user enters the longer string in lines 5-6. The string is stored in string variable **string1**.

In Lines 7-8, the program accepts the 2nd smaller string, and stores it in the variable **substring**.

In line 11, the program finds the position of the substring within the main string using the main function. The position is stored in an integer variable called **index**. If the substring is not found, it returns a value of **-1** for index.

The program then prints out the value index if it is greater than or equal to 0. If it is less than 0, then the program outputs that the substring was not found.

Output:

```
"C:\Users\abhis\Google Drive\C++\substring.exe"

Enter the main string: I won the game
Enter the substring to be found: won
The substring "won" was found at index 2 in the main string.

Process returned 0 (0x0)    execution time : 8.172 s
Press any key to continue.
```

In above example, substring **won** is found at index 2 in the main string **"I won the game"**.

```
"C:\Users\abhis\Google Drive\C++\substring.exe"

Enter the main string: I won the game
Enter the substring to be found: lost
The substring "lost" was not found in the main string.

Process returned 0 (0x0)    execution time : 4.873 s
Press any key to continue.
```

In above example output, substring **lost** is not found in the main string **"I won the game"**, so index is **-1**. Substring was not found.

Replace Strings

In C++, you can also replace certain words within a string. This is done using a function called **insert()**. In the code below, the program accepts two user inputs as string variables, **string1** and **substr**. **String1** is the larger string and **substr** is the string we want to insert within the string. The program then accepts an integer variable called **index**. This is the location within the string that we want to insert the substring.

```cpp
1    #include <iostream>
2    #include <string>
3
4    int main() {
5        std::string string1, substr;
6        int index;
7        std::cout << "Enter the main string: ";
8        std::getline(std::cin, string1);
9        std::cout << "Enter the substring to be inserted: ";
10       std::getline(std::cin, substr);
11       std::cout << "Enter the index location at which the substring should be inserted: ";
12       std::cin >> index;
13
14       string1.insert(index, substr);
15       std::cout << "The new string is: " << string1 << std::endl;
16
17       return 0;
18   }
19
```

In line 14, **substr** is inserted into **string1** at index location. The new string **string1** is printed out in line 15.

Output:

```
 "C:\Users\abhis\Google Drive\C++\strreplace.exe"
Enter the main string: Cakes, sugar and tea
Enter the substring to be inserted: the
Enter the index location at which the substring should be inserted: 14
The new string is: Cakes, sugar athend tea

Process returned 0 (0x0)   execution time : 57.674 s
Press any key to continue.
```

In the sample output, the main string string 1 is **"Cakes, sugar and tea"**. The substring is **"the"** When the substring is inserted at index 14, we get **"Cakes, sugar atthend tea"**.

Address

The below code is a good application of strings. Strings are used to get all elements of an address and print it out. The first line, second line, city, state, zip code and country are all entered as separate variables and printed on the screen. The inputs are entered in lines 6-17, and the outputs are printed out in lines 19-25.

A sample of the output is shown on the output below.

```cpp
#include <iostream>
#include <string>

int main() {
    std::string street1, street2, city, state, zip, country;
    std::cout << "Enter first line of the street address: ";
    std::getline(std::cin, street1);
    std::cout << "Enter second line of the street address: ";
    std::getline(std::cin, street2);
    std::cout << "Enter the city: ";
    std::getline(std::cin, city);
    std::cout << "Enter the state: ";
    std::getline(std::cin, state);
    std::cout << "Enter the country: ";
    std::getline(std::cin, country);
    std::cout << "Enter the zip code: ";
    std::getline(std::cin, zip);

    std::cout << "Address: " << std::endl;
    std::cout << street1 <<", " <<  std::endl;
    std::cout << street2 <<", " <<  std::endl;
    std::cout << city << ", " << std::endl;
    std::cout << state << " - ";
    std::cout << zip << ". " << std::endl;
    std::cout << country << std::endl;
    return 0;
}
```

Output:

```
"C:\Users\abhis\Google Drive\C++\address.exe"

Enter first line of the street address: 11
Enter second line of the street address: Smith Avenue
Enter the city: Chicago
Enter the state: Illinois
Enter the country: USA
Enter the zip code: 61122
Address:
11,
Smith Avenue,
Chicago,
Illinois - 61122.
USA

Process returned 0 (0x0)   execution time : 39.309 s
Press any key to continue.
```

Replace a Phrase

In C++, you can replace certain letters within a string with another string. This is done using the **replace()** function. In the below code, the program accepts three string variables as input. The first variable, **str**, is the main string. The second variable is called **phrase**, and it is the phrase that needs to be replaced. The third variable is called **new_phrase**, and it is the phrase that replaces the old one in **str**.

```
1    #include <iostream>
2    #include <string>
3
4    int main() {
5        std::string str, phrase, new_phrase;
6        std::cout << "Enter the main string: ";
7        std::getline(std::cin, str);
8        std::cout << "Enter the phrase to be replaced: ";
9        std::getline(std::cin, phrase);
10       std::cout << "Enter the new phrase: ";
11       std::getline(std::cin, new_phrase);
12
13       std::size_t index = str.find(phrase);
14       while (index <str.length()) {
15           str.replace(index, phrase.length(), new_phrase);
16           index = str.find(phrase, index + new_phrase.length());
17       }
18       std::cout << "The new string is: " << str << std::endl;
19
20       return 0;
21   }
22
```

In Line 13, the program checks if the phrase is located in **str** using the **find** function. It stores the index in an integer variable called **index**.

In Line 14, the program creates a while loop that runs while the **index** is smaller than the length of **str**. This is to ensure that all instances of phrase are found in **str**.

The first line in the loop in Line 15 replaces the **phrase** with **new_phrase** at the index. This **index** is the location of the phrase within str.

In Line 16, the program checks if there are any indexes within str that has the string variable phrase. If yes, the while loop continues till the entire string is checked.

Output:

```
 "C:\Users\abhis\Google Drive\C++\phrase_replace.exe"
Enter the main string: the fox ate the hen yesterday
Enter the phrase to be replaced: the
Enter the new phrase: a
The new string is: a fox ate a hen yesterday

Process returned 0 (0x0)   execution time : 18.697 s
Press any key to continue.
```

In the above sample output, str is **"the fox ate the hen yesterday"**. Phrase is **"the"**, and new phrase is **"a"**. So the program replaces every "the" with "a"; and we see that in the new output **"a fox ate a hen yesterday"**.

Problems to Practice

1. Write a program that accepts a 10-letter word from the user. If the length is 10, print out the words. Otherwise, print an error message.

2. Write a program that takes in 2 names as string variables and print out a sentence that combines the two names.

3. Write a program that takes in 5 string variables from a job candidate that store the candidate's name, occupation, zip code, salary and educational qualification. Print out all the information like a mini resume.

4. Write a program that takes in a sentence as an input and replaces all instances of curse words with ****

5. Write a program that finds the index of all instances of the word "help" in a sentence.

Conditionals

A conditional is a method to compare variables, and to make decisions in programming. It is a significant part of C++ programming. They are used to control the flow of the program and to respond to different situations.

The if else statement is the most commonly most conditional in C++. It's defined as below...

```
If (a==b)

{

//code 1

}

Else

{

//code 2

}
```

In the above code, it checks if the variable a is equal to variable b. If they are equal, then it executes code 1, else it executes code 2.

Let's look at a C++ example below.

If Else Example 1

```
1     #include <iostream>
2
3     int main() {
4         int a = 8;
5         int b = 15;
6         std::cout << " a = 8 "<< std::endl;
7         std::cout << " b = 15 "<< std::endl;
8         if (a > b) {
9             std::cout << "a is greater than b" << std::endl;
10        } else if (a < b) {
11            std::cout << "a is less than b" << std::endl;
12        } else {
13            std::cout << "a is equal to b" << std::endl;
14        }
15
16        return 0;
17    }
18
```

In line 4 and 5, it sets the values of variables a and b to 8 and 15 respectively. In lines 6 and 7, it prints the values on to the screen.

In line 8, it checks if the value of a is greater than b. If so, it executes the code in brackets between lines 8 and 10.

If a is less than b, it executes the code in brackets between 10 and 12.

If neither of the above two conditions are true, it means that a=b, and it executes the code in brackets between lines 12 and 14.

Output:

```
 "C:\Users\abhis\Google Drive\C++\conditionals_1.exe"

a = 8
b = 15
a is less than b

Process returned 0 (0x0)   execution time : 0.020 s
Press any key to continue.
```

In the output, it executes code between lines 8 and 10, as **a is less than b.**

If Else Example 1

In the next example, the program accepts value of two variables as inputs from the user; and it prints out the larger number using conditionals.

In lines 6,7 it accepts the value of the first number and stores it in integer variable **a**.

In lines 8,9 it accepts the value of the second number and stores it in integer variable **b**.

64

```
1      #include <iostream>
2
3      int main() {
4          int a, b;
5
6          std::cout << "Enter the first number: ";
7          std::cin >> a;
8
9          std::cout << "Enter the second number: ";
10         std::cin >> b;
11
12         if (a > b) {
13             std::cout << "The larger number is: " << a << std::endl;
14         } else if (a<b) {
15             std::cout << "The larger number is: " << b << std::endl;
16         } else {
17             std::cout << "Both numbers are equal" << std::endl;
18         }
19
20         return 0;
21     }
```

In line 12, it checks if the value of a is greater than b. If so, it executes the code in brackets between lines 12 and 14.

If a is less than b, it executes the code in brackets between 14 and 16.

If neither of the above two conditions are true, it means that a=b, and it executes the code in brackets between lines 16 and 18.

Output:

```
"C:\Users\abhis\Google Drive\C++\conditionals_2.exe"
Enter the first number: 10
Enter the second number: 8
The larger number is: 10

Process returned 0 (0x0)     execution time : 4.904 s
Press any key to continue.
```

Output:

```
 "C:\Users\abhis\Google Drive\C++\Math_Quiz_basic.exe"

Problem 1
What is 7 + 10? 17
Correct!
Problem 2
What is 14 * 3? 41
Incorrect. The correct answer is 42

Process returned 0 (0x0)    execution time : 4.488 s
Press any key to continue.
```

In output below, the user gets the first answer correct.

The second answer is wrong as he guesses 41 instead of 42.

Math Quiz with Random Numbers

The above program is modified to use random variables, and have the user guess the answer.

```
1     #include <iostream>
2     #include <cstdlib>
3     #include <ctime>
4
5     int main() {
6         int score = 0;
7         int sum1, sum2, diff1, diff2, answer;
8
9         sum1 = rand() % 100 + 1;
10        sum2 = rand() % 100 + 1;
11
12        std::cout << "What is " << sum1 << " + " << sum2 << "? ";
13        std::cin >> answer;
14
15        if (answer == sum1 + sum2)
16        {
17            std::cout << "Correct!" << std::endl;
18            score++;
19        } else {
20            std::cout << "Incorrect. The correct answer is " << sum1 + sum2 << std::endl;
21        }
22
23        diff1 = rand() % 100 + 1;
24        diff2 = rand() % 100 + 1;
25
26        std::cout << "What is " << diff1 << " - " << diff2 << "? ";
27        std::cin >> answer;
28
29        if (answer == diff1 - diff2)
30        {
31            std::cout << "Correct!" << std::endl;
32            score++;
33        } else {
34            std::cout << "Incorrect. The correct answer is " << diff1 - diff2 << std::endl;
35        }
36
37        return 0;
38    }
39
```

In this program, two random numbers between 1 to 100 are generated
and stored in variables **sum1** and **sum2**; as seen in lines 9-10.

The user is then asked to find the answer to the sum of the two above
variables, and the input is stored in integer variable **answer** in line 13.
An if condition is used to check if the **answer** is correct in line 15. If the
answer is correct, the program prints out "**Correct**", as in line 17.
Otherwise, it prints out "**Incorrect**" And notifies the user of the correct
answer is line 20.

The exact same process is done again with two new random integer variables **diff1** and **diff2**. This time the user is asked to input the difference of the two variables and the answer is checked again with another if condition.

Output:

```
"C:\Users\abhis\Google Drive\C++\Math_squiz.exe"
What is 42 + 68? 100
Incorrect. The correct answer is 110
What is 35 - 1? 34
Correct!

Process returned 0 (0x0)    execution time : 9.792 s
Press any key to continue.
```

In answer below, the two random variables sum1 and sum2 are 42 and 68. The user inputs 100 instead of 110 as the sum and so the user gets an incorrect answer.

In the second question, the two variables diff1 and diff2 are 35 and 1 respectively. The user inputs the correct answer of 34 and so gets a "Correct!" message.

Switch Case

Switch Case is another way of using conditionals in C++. It basically checks a variable's value using **switch** and then chooses among a bunch of **cases** based on the switch that happened. The program executes the **case** that is chosen.

It's defined as below.

```
Switch (var)

{

Case 1: //code A

Case 2: //code B

Case 3: //code C

}
```

In the above example, it executes code A if **var** is equal to 1; code B if **var** is equal to 2 and code C if **var** is equal to 3.

Switch Case Example

In below example, a user inputs the month as a number from 1-12. Based on the number input, it chooses a month using switch case.

In line 6-7, the user inputs a number, and it is stored in variable **month**.

In line 8, switch is applied to variable month.

Based on the input number, it executes code in corresponding case. For example, if variable is 3 it executes code between line 15 and 17. In these lines, string variable **monthName** is assigned to **March**. **break** in line 17 function exits the switch loop.

```cpp
#include <iostream>
#include <string>
int main() {
    int month;
    std::string monthName;
    std::cout << "Enter a number for the month (1-12): ";
    std::cin >> month;
    switch (month) {
        case 1:
            monthName = "January";
            break;
        case 2:
            monthName = "February";
            break;
        case 3:
            monthName = "March";
            break;
        case 4:
            monthName = "April";
            break;
        case 5:
            monthName = "May";
            break;
        case 6:
            monthName = "June";
            break;
        case 7:
            monthName = "July";
            break;
        case 8:
            monthName = "August";
            break;
```

```
33          case 9:
34              monthName = "September";
35              break;
36          case 10:
37              monthName = "October";
38              break;
39          case 11:
40              monthName = "November";
41              break;
42          case 12:
43              monthName = "December";
44              break;
45          default:
46              monthName = "Invalid month";
47      }
48
49      std::cout << monthName << std::endl;
50      return 0;
51  }
52
```

If numbers between 1-12 are not chosen, the code executed is default: which is between 45 and 47. MonthName is **"Invalid Month"**.

In lines 49 and 50, the code prints out the month name.

Output:

```
"C:\Users\abhis\Google Drive    ×    +  ∨

Enter a number for the month (1-12): 3
March

Process returned 0 (0x0)    execution time : 3.162 s
Press any key to continue.
```

In above example code, user chooses 3 and it executes code between line 15 and 17. string variable **monthName** is assigned to **March**. It prints out **March** as shown above.

Problems to Practice

1. Write a program takes in a number and checks if it's positive, negative or zero.
2. Write a program that takes in a number and checks if it's even or odd (check remainder when divided by 2).
3. Write a program that has a user choose between 10 brands of car at an auto dealership and prints out the brand and price based on the choice that the user makes.

Loops

A loop is a repetition of commands within a program a certain number of times. Every repetition is called an iteration of the loop.

Let's say you are looking to print out your name 1000 times. One option is to just write cout << "My name is xyz" and do it a 1000 times. That would take you at least about 10 minutes at least to copy and paste. It also takes a decent amount of memory space and is a lot of lines of codes.

The better option is to use loops as below:

```
for (int i=1; i<1000;i++)

{

std::cout<<"My name is xyz";

}
```

The loop is declared above with the counter variable **i=1**. It executes code between the brackets while **i** is less than 1000. And it increments by 1 each time; thus printing "**My name is xyz**" 1000 times.

Loops Example 1

In the first example below, the program prints out all numbers between 1 and 20.

```
1     #include <iostream>
2
3     int main() {
4         for (int i = 1; i <= 20; i++) {
5             std::cout << i << " "<< std::endl;
6         }
7         return 0;
8     }
9
```

In line 4, the loop is declared above with the counter variable i=1. It executes code between the brackets while i is less than and equal to 20. Between the brackets in lines 4-6, it prints the value of counter variable i. So, it prints all the values of i while it increments by 1 between 1 and 20.

This is seen in output below.

Output:

```
"C:\Users\abhis\Google Drive\C++\fornext.exe"
1
2
3
4
5
6
7
8
9
10
11
12
13
14
15
16
17
18
19
20

Process returned 0 (0x0)   execution time : 0.024 s
Press any key to continue.
```

Loops Example 2

In below example, the user is asked to enter a number and it prints all the numbers between 1 and that number. In lines 5-6, the user inputs an integer variable **looplength**.

```
1    #include <iostream>
2
3    int main() {
4        int looplength;
5        std::cout << "Enter how many numbers to print (starting at 1): ";
6        std::cin>>looplength;
7        for (int i = 1; i <= looplength; i++) {
8            std::cout << i << " "<< std::endl;
9        }
10       return 0;
11   }
12
```

In line 7, the loop integer variable **i** is set between 1 and **looplength** and increments by 1. Inside the loop, the program prints the value of loop integer variable **i**.

As seen in output below, the user inputs a **looplength** value of 15. So, the program prints integers between 1 and 15.

Output:

```
"C:\Users\abhis\Google Drive\C++\looplength.exe"

Enter how many numbers to print (starting 0): 15
1
2
3
4
5
6
7
8
9
10
11
12
13
14
15

Process returned 0 (0x0)    execution time : 2.941 s
Press any key to continue.
```

While Loop

The while condition is another type of loop which the code executes while a certain condition is met.

In example below, the loop code within the brackets executes while i is less than and equal to 20. The loop variable is declared outside the loop as **i=1**. It increments inside the loop while printing value of **i** inside as well.

```
1    #include <iostream>
2
3    int main() {
4        int i = 1;
5        while (i <= 20) {
6            std::cout << i << std::endl;
7            i++;
8        }
9        return 0;
10   }
11
```

As seen in output below, it prints all integers between 1 and 20.

Output:

```
"C:\Users\abhis\Google Drive\C++\whileloop.exe"

1
2
3
4
5
6
7
8
9
10
11
12
13
14
15
16
17
18
19
20

Process returned 0 (0x0)   execution time : 0.036 s
Press any key to continue.
```

Math Quiz with Random Numbers in a Loop

This program is similar to the "Math Quiz with Random Numbers" program in the **conditionals** chapter. Make sure that you understand the program before looking at this. The program basically accepts two random numbers as inputs and calculates their sum. The user guesses the answer; and the program prints an output based on the answer.

The program below is similar, but it uses a for loop to repeat the process 5 times. 5 different sets of random numbers are generated in lines 9-10 and stored in **num1** and **num2** during each iteration of the loop. An integer variable **score**, declared in line 7, keeps track of the correct answers by the user.

In lines 12-13, the user inputs an **answer** for their sum. In lines 15-21, the program checks if the sum is correct. If the answer is correct, it congratulates the user and **adds 1 to the score variable**. Otherwise, it tells the user what the correct answer.

It repeats the process 5 times in total and prints out user **score** (number of correct answers).

```
1     #include <iostream>
2     #include <cstdlib>
3     #include <ctime>
4
5     int main() {
6         int num1, num2, answer;
7         int score = 0;
8         for (int i = 0; i < 5; i++) {
9             num1 = rand() % 100 + 1;
10            num2 = rand() % 100 + 1;
11
12            std::cout << "What is " << num1 << " + " << num2 << "? ";
13            std::cin >> answer;
14
15            if (answer == num1 + num2)
16            {
17                std::cout << "Correct!" << std::endl;
18                score++;
19            } else {
20                std::cout << "Incorrect. The correct answer is " << num1 + num2 << std::endl;
21            }
22
23        }
24        std::cout << "Your final score is " << score << "/5" << std::endl;
25     return 0;
26     }
27
```

Output:

```
"C:\Users\abhis\Google Drive\C++\Math_Quiz_Random_Loop.exe"

What is 42 + 68? 110
Correct!
What is 35 + 1? 36
Correct!
What is 70 + 25? 95
Correct!
What is 79 + 59? 138
Correct!
What is 63 + 65? 128
Correct!
Your final score is 5/5

Process returned 0 (0x0)    execution time : 13.823 s
Press any key to continue.
```

In the above example, the user gets all correct answers, so the score variable gets to 5 and the final score is 5 out of 5.

Problems to Practice

1. Write a program that takes in two random numbers and has a user guess their product. Print out whether the user got the answer correct. Do this 5 times in a FOR loop and print out how many correct answers the user got.

2. Write a program that takes in two random whole numbers and figures out which number is higher. Divide the higher number by the lower number and have the user guess the remainder. Print out whether the user got the answer correct. Do this 5 times in a FOR loop and print out how many correct answers the user got.

3. Write a guessing game where the user has to guess a randomly generated number between 1 and 100. If the number is too high, say "too high", otherwise say it's "too low". Keep doing this till the user guesses correctly (Hint: use a while loop).

Timers and Clocks

Timers and clocks in C++ are mechanisms to find the amount of elapsed time between the start and end of a certain event.

This is great for performance analysis, gaming, animations, alarm clocks and optimization.

Several different libraries are used for different types of timers, such as <unistd>,<ctime>,<chrono> etc.

Countdown Clock

In the first example, we'll go through the **sleep()** function which is a sort of countdown clock. The sleep function pauses the program for a certain number of seconds.

The sleep function is part of the **<unistd.h>** library, which is invoked in Line 2 of below program.

The program below, in lines 6-7, accepts an integer variable **timetoalarm**, which holds the number of seconds the program is to pause.

In Line 9, the program invokes the sleep function. It pauses the program for **timetoalarm** seconds.

In Line 10, the program prints an output to the program which indicates that the sleep function has elapsed.

```
1    #include <iostream>
2    #include <unistd.h>
3
4    int main() {
5        int timetoalarm;
6        std::cout << "Enter the time for the alert (in seconds): ";
7        std::cin >> timetoalarm;
8        std::cout << "Timer set for " << timetoalarm << " seconds." << std::endl;
9        sleep(timetoalarm);
10       std::cout << timetoalarm<< " seconds over!" << std::endl;
11       return 0;
12   }
13
```

Output:

In the code below, you will notice a pause of 6 seconds before the text "6 seconds over!" is printed on the screen.

```
"C:\Users\abhis\Google Drive\C++\Timer.exe"
Enter the time for the alert (in seconds): 6
Timer set for 6 seconds.
6 seconds over!

Process returned 0 (0x0)    execution time : 8.835 s
Press any key to continue.
```

Get Current Date and Time

In C++, we can get the information about the current date and time using the **time** and **localtime** function.

The code below prints today's date and time to the output screen. It uses the time function from the **<ctime library>** which is invoked in line 2.

In line 5, a **time_t** variable called **cTime** is assigned to the current time using the time function.

In line 6, the **localtime** function converts **cTime** into a **localtime** variable. This broken down **localtime** variable has year, month, day information that can be printed on the screen.

In lines 8-14, the different components of the **localtime** variable are printed on the screen. (Note that the year needs to have 1900 added to it, as the program only detects how many years past 1900 have passed. For example, if the date is 2020; the program would output 120. We need to add 1900 to ensure the correct date. For the same reason, the number 1 needs to be added to the month.)

```
1    #include <iostream>
2    #include <ctime>
3
4    int main() {
5        std::time_t cTime = std::time(nullptr);
6        std::tm* localtime = std::localtime(&cTime);
7
8        std::cout << "Today's date and time:" << std::endl;
9        std::cout << "Year: " << localtime->tm_year + 1900 << std::endl;
10       std::cout << "Month: " << localtime->tm_mon + 1 << std::endl;
11       std::cout << "Day: " << localtime->tm_mday << std::endl;
12       std::cout << "Hour: " << localtime->tm_hour << std::endl;
13       std::cout << "Minute: " << localtime->tm_min << std::endl;
14       std::cout << "Second: " << localtime->tm_sec << std::endl;
15
16       return 0;
17   }
18
```

Output:

```
"C:\Users\abhis\Google Drive\C++\currentTime.exe"

Today's date and time:
Year: 2023
Month: 1
Day: 20
Hour: 16
Minute: 7
Second: 20

Process returned 0 (0x0)   execution time : 0.060 s
Press any key to continue.
```

95

Alarm clock

An alarm clock is set to activate an alert at a particular date and time. There is no specific alarm clock in C++. However, in the program below, we use the **localtime** and **sleep** function to make an alarm clock. We use the **localtime** function to determine the difference in time between current time and alarm clock time; and then we use the sleep function to pause the program till alarm time.

The program uses functions from **<unistd.h>** and **<ctime>** which are invoked in lines 2-3 of the program.

In lines 7-10, the program accepts user input for hour and minute of time. In this program, we are doing an alarm within the same day as we don't want to keep the program running for multiple days.

In Lines 13-17, **time** and **localtime** functions are used to get the current time and stores it in a time variable called **currentTime**.

In Line 20, a new time variable called **alarmTime** is created which stores the time of the alarm.

In line 21, the program calculates the difference between **alarmTime** and **localtime** and stores it in a time variable called **timeUntilAlarm**.

In Line 26, the line uses the **sleep** function to pause the program till the alarm clock time.

In line 28, the program prints an output to the screen which indicates that it's time for the alarm.

```cpp
1    #include <iostream>
2    #include <ctime>
3    #include <unistd.h>
4
5    int main() {
6        int ahour, aminute;
7        std::cout << "Enter the hour for the alarm (between 0-23): ";
8        std::cin >> ahour;
9        std::cout << "Enter the minute for the alarm (between 0-59): ";
10       std::cin >> aminute;
11
12       // Get current time
13       std::time_t currentTime = std::time(nullptr);
14       std::tm* localTime = std::localtime(&currentTime);
15       localTime->tm_hour = ahour;
16       localTime->tm_min = aminute;
17       localTime->tm_sec = 0;
18
19       // Calculate time until alarm
20       std::time_t alarmTime = std::mktime(localTime);
21       std::time_t timeUntilAlarm = alarmTime - currentTime;
22
23       std::cout << "Alarm set for " << ahour << ":" << aminute << "." << std::endl;
24       std::cout << "Time until alarm: " << timeUntilAlarm << " seconds." << std::endl;
25
26       sleep(timeUntilAlarm);
27
28       std::cout << "It's alarm time!" << std::endl;
29       return 0;
30   }
31
```

Output:

```
 "C:\Users\abhis\Google Drive\C++\alarmclock.exe"

Enter the hour for the alarm (between 0-23): 16
Enter the minute for the alarm (between 0-59): 15
Alarm set for 16:15.
Time until alarm: 0 seconds.
It's alarm time!

Process returned 0 (0x0)   execution time : 4.976 s
Press any key to continue.
```

97

Stopwatch

A stopwatch in C++ is used to determine the amount of time elapsed between two events. Both events occur by pressing "Enter" on the keyboard. The stopwatch uses a function called **cin.get()** to detect when "Enter" is pressed on the keyword, to start and stop the clock.

It records the start time using **chrono::high_resolution_clock::now()** the first time the user presses enter. The stopwatch is now running. Now, the user presses enter again to stop the clock. The program records the end time. It computes the elapsed time between the start and end times as a duration in milliseconds using **chrono::duration_cast<std::chrono::milliseconds>(end - begin)**.

The output from chrono function is in milliseconds, so it has to be divided by 1000 before being output on the screen.

```cpp
1    #include <iostream>
2    #include <chrono>
3
4    int main() {
5        std::cout << "Press Enter to Start the Clock." << std::endl;
6        std::cin.get();
7
8        auto begin = std::chrono::high_resolution_clock::now();
9
10       std::cout << "Press Enter to stop the stopwatch." << std::endl;
11       std::cin.get();
12
13       auto end = std::chrono::high_resolution_clock::now();
14       auto tme = std::chrono::duration_cast<std::chrono::milliseconds>(end - begin);
15
16       std::cout << "Elapsed time: " << tme.count()/1000 << " seconds." << std::endl;
17       return 0;
18   }
19
```

Doing Math with C++

The high processing capabilities of C++ make it the ideal programming language to complete all kinds of computations, from kindergarten Math problems to industry level computations.

In this chapter, we'll go over a few commonly used C++ internal math functions; and a few others that are not well known but very useful as well.

In all the programs in this chapter, we invoke the **cmath** library by using the line: **#include <cmath>**

Round Numbers

C++ has the ability to round decimal numbers to the nearest whole number. We use a function called **round()** function to achieve this.

This is shown in the example below:

```
1    #include <iostream>
2    #include <cmath>
3
4
5    int main() {
6        float dnum;
7        std::cout << "Enter a decimal number: ";
8        std::cin >> dnum;
9
10       std::cout << "Rounded to nearest whole number: " << round(dnum) << std::endl;
11
12       return 0;
13   }
14
```

In Line 2, the **cmath** library which contains all C++ Math functions is invoked.

In Line 7, the program declares a float variable called **dnum** which holds the decimal number value. A **float** variable is one which can hold decimal values. An integer variable can only hold whole number values.

In Lines 8 and 9, the program accepts a decimal number from the user and stores it in variable **dnum**.

In Line 11, the decimal number is rounded to the nearest whole number using **round(dnum)** and display it on the screen using **cout** command.

104

We can see the outputs below, for 2 instances of running the program. In the first instance, 5.45 gets rounded down to 5. In the second instance, 3.67 gets rounded to the closest whole number of 4.

Output:

```
"C:\Users\abhis\Google Drive\C++\roundnumber.exe"

Enter a decimal number: 5.45
Rounded to nearest whole number: 5

Process returned 0 (0x0)    execution time : 6.279 s
Press any key to continue.
```

```
"C:\Users\abhis\Google Drive\C++\roundnumber.exe"

Enter a decimal number: 3.67
Rounded to nearest whole number: 4

Process returned 0 (0x0)    execution time : 4.437 s
Press any key to continue.
```

Round Decimals Up and Down

C++ also has the ability to individually round decimals up or down. **Ceil()** rounds a decimal to the higher whole number, while **floor()** rounds a decimal down to a lower whole number.

The below program takes in a decimal number and stores it in a variable dnum. It calculates and prints out both the rounded up and rounded down whole number. As seen in the output below the code, 10.83 is rounded up to 11 and rounded down to 10.

```
1    #include <iostream>
2    #include <cmath>
3
4
5
6    int main() {
7        float dnum;
8        std::cout << "Enter a decimal number: ";
9        std::cin >> dnum;
10
11       std::cout << "Decimal Number Rounded up: " << ceil(dnum) << std::endl;
12       std::cout << "Rounded down: " << floor(dnum) << std::endl;
13
14       return 0;
15   }
16
```

Output:

```
"C:\Users\abhis\Google Drive\C++\roundupdown.exe"

Enter a decimal number: 10.83
Decimal Number Rounded up: 11
Rounded down: 10

Process returned 0 (0x0)    execution time : 6.560 s
Press any key to continue.
```

Calculate remainder

In C++, we calculate the reminder using a **modulus** operator which is represented by the **%** sign below. So, if 21 divided by 5 has remainder of 1, in C++ it would be:

21 % 5 = 1

In below example, the program takes in two integer variables **n1** and **n2** and finds the reminder when **n1 is divided by n2**.

In Line 10, the remainder is calculated using **n1%n2** and the output is printed on the screen.

```
1    #include <iostream>
2
3
4    int main() {
5        int n1, n2;
6        std::cout << "Enter numerator: ";
7        std::cin >> n1;
8        std::cout << "Enter denominator: ";
9        std::cin >> n2;
10       std::cout << "The remainder when "<< n1 << " is divided by "<< n2 << " is = " << n1 % n2 << std::endl;
11       return 0;
12   }
13
```

Output:

```
 "C:\Users\abhis\Google Drive\C++\remainder.exe"
Enter numerator: 17
Enter denominator: 3
The remainder when 17 is divided by 3 is = 2

Process returned 0 (0x0)   execution time : 17.197 s
Press any key to continue.
```

In above example, n1 is 17 and n2 is 3 and the remainder when 17 is

divided by 3 is 2, as shown in output.

```
 "C:\Users\abhis\Google Drive\C++\remainder.exe"
Enter numerator: 45
Enter denominator: 6
The remainder when 45 is divided by 6 is = 3

Process returned 0 (0x0)   execution time : 11.069 s
Press any key to continue.
```

In above example, n1 is 45 and n2 is 6 and the remainder when 45 is

divided by 6 is 3, as shown in above output.

Basic Math Operations

Basic arithmetic operations in C++ are conducted using their arithmetic sign.

Addition is done using "+"

Subtraction is done using "-"

Multiplication is done using "*"

Division is done using "/"

In below code, the program accepts two float variables **n1** and **n2**; and calculates their sum, difference, product and difference.

The math operations are done in lines 10-13 and the output is printed on the screen.

```
1    #include <iostream>
2
3
4    int main() {
5        float n1, n2;
6        std::cout << "Enter first number: ";
7        std::cin >> n1;
8        std::cout << "Enter second number: ";
9        std::cin >> n2;
10       std::cout << n1 << " + " << n2 << " = " << n1 + n2 << std::endl;
11       std::cout << n1 << " - " << n2 << " = " << n1 - n2 << std::endl;
12       std::cout << n1 << " * " << n2 << " = " << n1 * n2 << std::endl;
13       std::cout << n1 << " / " << n2 << " = " << n1 / n2 << std::endl;
14       return 0;
15   }
16
```

Output:

```
"C:\Users\abhis\Google Drive    ×    +    ⌄

Enter first number: 5
Enter second number: 7
5 + 7 = 12
5 - 7 = -2
5 * 7 = 35
5 / 7 = 0.714286

Process returned 0 (0x0)    execution time : 7.285 s
Press any key to continue.
```

In above example, the two integers n1 and n2 are 5 and 7.

The outputs when basic math operations are performed are shown above.

Exponent

In C++, an exponent or power is calculated using the **pow()** function. For example, if we want to calculate 2 raised to the power 5, we use:

pow(2,5) which gives an output of **32**.

In below code, the program accepts two float variables called **base** and **exp**. The **base** is the base of the exponent and **exp** is the power to which the base is raised.

In line 11, the program calculates the power of **base** raised to **exp** and stores it in a third float variable called **res**.

In line 12, the output **res** is printed on the screen.

```
1    #include <iostream>
2    #include <cmath>
3
4    int main() {
5        float base, exp;
6        std::cout << "Enter the base number: ";
7        std::cin >> base;
8        std::cout << "Enter the exponent: ";
9        std::cin >> exp;
10
11       float res = pow(base, exp);
12       std::cout <<base<<" raised to the power "<<exp<< " is " << res << std::endl;
13
14       return 0;
15   }
16
```

Output:

```
 "C:\Users\abhis\Google Drive\C++\exponent.exe"
Enter the base number: 5
Enter the exponent: 2
5 raised to the power 2 is 25

Process returned 0 (0x0)    execution time : 9.630 s
Press any key to continue.
```

In above example, **base** is 5 and **pow** is 2. The output variable **res** is 5 raised to the power of 2, which is 25.

Logarithm

In C++, a logarithm is calculated using the **log()** function. There is no way to include base in the logarithm in C++. So the log of a number has to be divided by the log of its base. For example, the log of 20 to the base 2 is given by:

log (20) / log (2)

In the below code, the program accepts two float variables **num** and **base** in lines 6-9. The base is given by the variable **base**.

In line 11, the program calculates the log of num and stores it in a float variable called **logn**.

In line 12, the output is printed on the screen.

```
1   #include <iostream>
2   #include <cmath>
3
4   int main() {
5       float num, base;
6       std::cout << "Enter the number: ";
7       std::cin >> num;
8       std::cout << "Enter the log base: ";
9       std::cin >> base;
10
11      float logn = log(num) / log(base);
12      std::cout <<" The logarithm of "<<num<<" to the base "<<base<< " is " << logn << std::endl;
13
14      return 0;
15  }
16
```

Output:

```
■ "C:\Users\abhis\Google Drive\C++\log.exe"
Enter the number: 100
Enter the log base: 10
 The logarithm of100 to the base 10 is 2

Process returned 0 (0x0)   execution time : 2.748 s
Press any key to continue.
```

In above output, **num** is 100 and **base** is 10. The output **logn** calculates

the log, which is 2.

Square root

In C++, the square root of a number is calculated using the **sqrt()** function.

In the below code, the program accepts a float variable called **num**.

In line 9, the program calculates the square root of num and stores it in a float variable called **res**.

In line 10, the output is printed on the screen.

```cpp
1    #include <iostream>
2    #include <cmath>
3
4    int main() {
5        float num;
6        std::cout << "Enter the number: ";
7        std::cin >> num;
8
9        float res = sqrt(num);
10       std::cout <<" The square root of "<<num<< " is " << res << std::endl;
11
12       return 0;
13   }
14
```

Output:

```
 ■ "C:\Users\abhis\Google Drive\C++\squareroot.exe"
Enter the number: 36
 The square root of 36 is 6

Process returned 0 (0x0)    execution time : 6.520 s
Press any key to continue.
```

In above example, the variable **num** is 36 and **res** is 6, which is output to the screen.

Trigonometry

C++ has a wide range of trigonometrical functions. In this book, we'll go over a few basic functions. The sine, cosine and tangent of an angle is given by **sin()**, **cos()** and **tan()** functions respectively. The angle must be in radians. For example, if we wanted to calculate sine of 30 degrees, we use:

sin (30*M_PI/180) instead of **sin (30)**

where M_PI is 3.142 in C++

In code below, the program accepts the angle in degrees and stores it in a float variable called **angle**.

In line 8, the value of the angle is converted to radians and stored in a double variable called **angle_radian**.

In line 9-11, the program calculates the values of sine, cosine and tangent of angle and stores the values in double variables **sine**, **cosine** and **tangent**.

In lines 12-14, the values are output on the screen.

```
1    #include <iostream>
2    #include <cmath>
3
4    int main() {
5        float angle;
6        std::cout << "Enter the angle (in degrees): ";
7        std::cin >> angle;
8        double angle_radian = angle*M_PI/180;
9        float sine = sin(angle_radian);
10       float cosine = cos(angle_radian);
11       float tangent = tan(angle_radian);
12       std::cout <<" The sine of "<<angle<< " is " << sine << std::endl;
13       std::cout <<" The cosine of "<<angle<< " is " << cosine << std::endl;
14       std::cout <<" The tangent of "<<angle<< " is " << tangent << std::endl;
15       return 0;
16   }
17
```

Output:

```
"C:\Users\abhis\Google Drive\C++\sincostan.exe"

Enter the angle (in degrees): 30
 The sine of 30 is 0.5
 The cosine of 30 is 0.866025
 The tangent of 30 is 0.57735

Process returned 0 (0x0)    execution time : 3.284 s
Press any key to continue.
```

In above code, the user inputs a value of 30, and the program

calculates and prints out the sine, cosine and tangent of 30 degrees.

Hypotenuse

The hypotenuse of a right-angled triangle is calculated using the square root function **sqrt()**. The Sqrt function is used on the sum of the squares of the two sides of the right angle.

In below code, the program accepts two float variables **side1** and **side2**. These are the two sides of the right angle.

The hypotenuse is stored in a third float variable and is stored in a variable called **hypotenuse**.

In line 12, the value of the hypotenuse is calculated and stored in **hypotenuse** variable.

In line 13, this is output onto the screen.

```
1   #include <iostream>
2   #include <cmath>
3
4   int main() {
5       float side1, side2, hypotenuse;
6       std::cout << "Enter the base of the right angled triangle: ";
7       std::cin >> side1;
8       std::cout << "Enter the vertical side of the right angled triangle: ";
9       std::cin >> side2;
10
11
12      hypotenuse = sqrt(side1*side1+side2*side2);
13      std::cout <<" The hypotenuse of the triangle is " << hypotenuse << std::endl;
14
15      return 0;
16  }
17
```

Output:

```
Enter the base of the right angled triangle: 5
Enter the vertical side of the right angled triangle: 12
 The hypotenuse of the triangle is 13

Process returned 0 (0x0)    execution time : 3.533 s
Press any key to continue.
```

In above sample code, **side1** and **side2** are 5 and 12.

The hypotenuse is calculated as square root of 5 squared plus 12 squared, which comes to an answer of 13.

Problems to Practice

1. Write a program that takes in 3 numbers as integer values; and calculates their sum, difference, product, and quotient.

2. Write a program that takes in 2 numbers and finds the quotient and remainder when 1st number is divided by the 2nd number.

3. Write a program that takes in the radius of a circle as a float variable and finds the area of the circle ($A=pi*r^2$).

4. Write a program that takes in an angle and one side of a right-angled triangle as two float variables. Calculate the values of the two missing sides. (Use sine and cos values of angle)

5. Write a program that takes in an angle and hypotenuse of a right-angled triangle as two float variables. Calculate the values of the two missing sides. (Use sine and cos values of angle)

6. Write a program that takes in a number as a float variable and finds the cube of the number. (Use an exponent of 3 for cube)

7. Write a program that takes in a number as a float variable and find the cube root. (Use an exponent of 1/3 for cube root)

8. Write a program that takes in the value of a hypotenuse and one side of a right-angled triangle. The program finds the value of the missing side and prints it out.

Arrays

An array is a collection of elements of a particular data type. Each element is indexed and can be accessed by using its index. It has a fixed number of elements that is defined by the user.

Arrays are great tools to manipulate and store data effectively, as they are great for fast access to data and iteration.

In C++, arrays are defined as below:

int arr[6]={10,20,30,40,50,60};

arr is an integer array of 6 elements shown above.

The first index in a C++ array is always 0.

So, in the above example, if we want to change the second element to 15:

arr[1]=15;

Then the array arr becomes:

arr[6] = {10,15,30,40,50,60}

Basic Array

In our first example, we create an array of 15 elements and print out all numbers from 0-14 using the indexes of the array.

In the code below, the array **arr** is defined in line 4 as an integer array of 15 elements.

In lines 5-7, each element of the array is assigned as a number from **0 to 14** in ascending order. This is done using a for loop where the loop indexes from 0 to 14.

In lines 8-10, the elements of the array are printed out using a for loop. This is shown in output below the code.

```
1    #include <iostream>
2
3    int main() {
4        int arr[15];
5        for (int i = 0; i < 15; i++) {
6            arr[i] = i;
7        }
8        for (int i = 0; i < 15; i++) {
9            std::cout << "Array Element "<<i<<" = "<<arr[i] << std::endl;
10       }
11       return 0;
12   }
13
```

```
1    #include <iostream>
2
3    int main() {
4        int sizeofArray;
5        std::cout << "Enter the size of the array: ";
6        std::cin >> sizeofArray;
7
8        int myarr[sizeofArray];
9
10       for (int i = 0; i < sizeofArray; i++) {
11           std::cout << "Enter element " << i+1 << ": ";
12           std::cin >> myarr[i];
13       }
14
15       std::cout << "The elements of the array are: ";
16       for (int i = 0; i < sizeofArray; i++) {
17           std::cout << myarr[i] << " ";
18       }
19
20       return 0;
21   }
22
```

Output:

```
"C:\Users\abhis\Google Drive\C++\Basicarray2.exe"

Enter the size of the array: 5
Enter element 1: 34
Enter element 2: 23
Enter element 3: 45
Enter element 4: 11
Enter element 5: 4
The elements of the array are: 34 23 45 11 4
Process returned 0 (0x0)    execution time : 10.261 s
Press any key to continue.
```

In output above, the size of the array is entered as 5, which creates an array **myArr[5]**. The 5 elements of **arr** are 34,23,45,11 and 4.

130

The array is then printed on the screen.

Reverse Name

In C++, a string is just an array of characters. Each element of the string is a character and can be accessed by the index. For example,

string name = "Tim"

name is an array of 3 characters. So, name[0]="T".

If we do name[1]="o", then:

name = "Tom"

In below program, we use this fact to reverse a string array. The string array is a name input by the user.

In line 7, the program accepts a name from the user and stores it in a string array called **name**.

In line 11-13, the program has a for loop that goes from name.length-1 to 0. So, if the name is "tommy", name.length is 5. So, the for loop goes from 4 to 0. It runs 5 times in total.

In line 12, **cout<<name[i]** prints the character of the array that is based on the loop index.

The loop, in effect, prints out the name in reverse.

```
1       #include <iostream>
2       #include <string>
3
4    ☐int main() {
5           std::string name;
6           std::cout << "Enter your first name: ";
7           std::cin >> name;
8           std::cout << std::endl;
9
10          std::cout << "Name reversed is ";
11   ☐      for (int i = name.length() - 1; i >= 0; i--) {
12              std::cout << name[i];
13   ├       }
14          std::cout << std::endl;
15
16          return 0;
17   └ }
18
```

Output:

In the output below, **"tommy"** is entered by the user. The program

prints out **"ymmot"**, which is the reverse of the string.

```
#include <iostream>
 "C:\Users\abhis\Google Drive\C++\reversename.exe"
Enter your first name: tommy

Name reversed is ymmot

Process returned 0 (0x0)   execution time : 12.033 s
Press any key to continue.
```

Guess US State Capitals

In below example, we play a game of US Capitals. The user is prompted with a question on the capital of a country. Based on the output, the program either prints out that the user is correct; or prints the correct answer.

The program uses two arrays, called **states[]** and **capitals[]**, which are defined in lines 9 and 10.

Both these arrays are in the **appendix** of this book and can be copied and pasted into the program.

In line 12, the program determines the size of the array **states[]** and stores in an integer variable called **numStates**.

In line 14, two string variables **state** and **correctedCapital** are created. **state** holds the value of a random state that the program gets from the array; and **correctedCapital** is the capital of that **state**. The program basically gets a random state from the array using the **rand()** function. To invoke the rand function, we also invoke the random library in line 3.

Since we want to run this quiz 5 times, the program runs a for loop 5 times in line 15.

Inside the loop in line 16, a random number is generated between 1 and **numStates** and stored in integer variable **randomIndex**.

In line 17, the state based on **randomIndex** is generated and stored in string variable **state**. In line 18, the capital of state is stored in string variable **correctedCapital**.

In line 20, the user enters his guess for the capital of state and the answer is stored in string variable **userGuess**.

An if loop is used to check if **userGuess** is correct. If it correct, a **"Correct"** message is printed on the screen. An integer variable **score** is increased by 1. The score variable keeps track of the number of correct answers by the user.

Otherwise, an incorrect message with the correct answer is printed on the screen.

This sequence is run 5 times until the for loop is over.

In line 30, the program the number of correct answers by the user.

```cpp
1    #include <iostream>
2    #include <string>
3    #include <random>
4
5
6    int main() {
7        int score = 0;
8        std::string states[] = {"Alabama","Alaska","Arizona","Arkansas","California","Colorad
9        std::string capitals[] = {"Montgomery","Juneau","Phoenix","Little Rock","Sacramento",
10
11
12       int numStates = sizeof(states) / sizeof(states[0]);
13
14       std::string state, correctCapital;
15       for (int i = 0; i < 5; i++) { // play 10 rounds
16           int randomIndex = rand() % numStates + 1;
17           state = states[randomIndex];
18           correctCapital = capitals[randomIndex];
19           std::cout << "What is the capital of " << state << "? ";
20           std::string userGuess;
21           std::cin >> userGuess;
22           if (userGuess == correctCapital) {
23               std::cout << "Correct!\n";
24               score++;
25           } else {
26               std::cout << "Incorrect. The correct answer is " << correctCapital << ".\n";
27           }
28       }
29
30       std::cout << "Your final score is " << score << " out of 5.\n";
31
32       return 0;
33   }
34
```

135

Output:

```
#include <iostream>
■ "C:\Users\abhis\Google Drive\C++\WorldCapitals.exe"
What is the capital of Texas? Austin
Correct!
What is the capital of Maine? idk
Incorrect. The correct answer is Augusta.
What is the capital of Oklahoma? Oklahoma City
Incorrect. The correct answer is Oklahoma City.
What is the capital of Alaska? Incorrect. The correct answer is Juneau.
What is the capital of Massachusetts? Boston
Correct!
Your final score is 2 out of 5.

Process returned 0 (0x0)   execution time : 21.106 s
Press any key to continue.
```

In above output, the user is asked 5 questions on capitals of US states.

He answers 5 but only gets 2 of them correct, which are the capitals of Texas and Oklahoma City.

Problems to Practice

1. Write a program that creates a string array of 5 names. It then asks the user to enter 5 names into the array. The program then prints out the final array of 5 names.

2. A class has 20 students who just completed a Physics test. Create two arrays, one a string array that contains names, and another one which is an integer array that contains Physics marks. The program prompts the user to enter names and Physics marks of all students. At the end, the program prints out the average score, the highest score and the person with the highest score.

3. Write a program that takes in an array and creates a new array of half the length with every odd character. It then prints out the new array.

4. Write a program that creates an array of world countries and capitals. Create a game where it asks the user 5 times the capital of a certain country and adds the number of times the user guesses correct. At the end, it prints out the number of times the user guessed right.

Pointers

A pointer is a variable that holds the memory location of another variables. It is a special feature of C and C++, though a few other languages hold it.

Let's say we have a variable x, and we define that as equal to 10 in C++ using:

int var = 10;

var has a value of 10 but it is assigned to a particular space in the computer memory. To access this value, we use:

int* pvar = &var;

pvar holds the memory location so if we print **pvar** on the screen we get something like **0c76fe89.**

If we only know the name of the pointer but not the variable, we can use ***pvar** to get the value of what is stored in the pointer location. In this case, both ***pvar** and **var** have a value of 10.

Pointers are great because they allow for efficient use of computer memory and allow for faster computer processing. While we won't go into the advanced programs that pointers are used for, we'll go over the basics of pointers with some simple examples so we can see how they are used.

The main difference between a variable and a pointer is that if the value of the variable changes, the value of the pointer remains the same. For example, if we do:

x=20;

the value of **pvar** still remains 0c76fe89. We can also change the value of x by changing the value of *****pvar.**

*pvar =30;

Means that the value of x also changes to 30.

Pointer Basic Example 1

A pointer is a variable that holds the memory location of another variable.

In the below example, **num** is the variable and is assigned to a value of 20 in line 4.

In line 5, a pointer variable **pnum** is assigned the memory location of **num**.

In line 6, 7 and 8; the program prints the value of **num**, **pnum** and ***pnum** respectively. As seen in output below the code, **pnum** has the memory location of num, while ***pnum** holds the value of **num**.

In line 9, the program uses ***pnum** to change the value of the variable to 100. This also changes the value of **num** as shown in the output below the code.

```cpp
1    #include <iostream>
2
3    int main() {
4            int num = 20;
5            int* pnum = &num;
6            std::cout << "num = " << num << std::endl;
7            std::cout << "pnum = " << pnum << std::endl;
8            std::cout << "*pnum = " << *pnum << std::endl;
9            *pnum = 100;
10           std::cout << "num = " << num << std::endl;
11           return 0;
12   }
13
```

Output:

In the beginning num has a value of 20, while **pnum** has a value of

0x61f214. ***pnum** holds the value of num, so it prints out 20.

After the program assigns ***pnum** a value of 100, we see that **num** is

changed to 100.

```
"C:\Users\abhis\Google Drive\C++\Pointerexample.exe"
num = 20
pnum = 0x61fe14
*pnum = 20
num = 100

Process returned 0 (0x0)   execution time : 0.059 s
Press any key to continue.
```

Making it Point to a Different Variable

A common strategy in high computation problems is to use the same pointer for multiple variables. This makes the program run faster as it uses less memory.

In below code, the program uses the same pointer for multiple variables. The two variables are **num1** and **num2** assigned to values of **20** and **40** respectively.

In line 6, the pointer first points to **num1**. In lines 7-10, the values of **num1**, **num2**, **pnum** and ***pnum** are printed. **pnum** holds the memory address, while ***pnum** holds the value of **num1**, which is **20**.

Then, in line 12, the same pointer **pnum** points to **num2**. In lines 13-17, the same values are printed again. This time, ***pnum** is now 40 as that is the value of num2. But we notice that the memory location **pnum** is the same. It is the same memory address, but it now points to num2.

This is shown in the output below.

```cpp
1    #include <iostream>
2
3    int main() {
4        int num1 = 20;
5        int num2 = 40;
6        int* pnum = &num1;
7        std::cout << "num1 = " << num1 << std::endl;
8        std::cout << "num2 = " << num2 << std::endl;
9        std::cout << "pnum = " << pnum << std::endl;
10       std::cout << "*pnum = " << *pnum << std::endl;
11
12       pnum=&num2;
13       std::cout << "Now pnum points to num2" << std::endl;
14       std::cout << "num1 = " << num1 << std::endl;
15       std::cout << "num2 = " << num2 << std::endl;
16       std::cout << "pnum = " << pnum << std::endl;
17       std::cout << "*pnum = " << *pnum << std::endl;
18
19
20       return 0;
21   }
22
```

Output:

```
"C:\Users\abhis\Google Drive\C++\Pointerexample2.exe"

num1 = 20
num2 = 40
pnum = 0x61fe14
*pnum = 20
Now pnum points to num2
num1 = 20
num2 = 40
pnum = 0x61fe10
*pnum = 40

Process returned 0 (0x0)   execution time : 0.056 s
Press any key to continue.
```

Pointers to Swap Variables

Swapping variables is a good way to utilize pointers. Instead of changing memory locations, the same memory locations can be used which allows for faster processing.

In the below code, two integer variables **x** and **y** have values of 50 and 25 before swapping as output in line 11.

The swapping process occurs in lines 3-6 in the function called **swap**. The function accepts the memory values of x and y as shown in line 13. **&x** and **&y** are the memory locations.

In line 4, the value of a (represented by a*) is sent to a temporary variable called **temp**.

In line 5, *a is assigned to *b. This means that the memory location that used to hold the value of a now holds the value of b.

In line 6, *b holds the value of temp. This means the memory location of b now holds temp, which is the value of a.

So, the values that a and b hold are swapped, while keeping the memory locations the same.

This is shown when the values are printed on the screen in line 15 in the output below the code.

```cpp
1    #include <iostream>
2
3    void swap(int* a, int* b) {
4        int temp = *a;
5        *a = *b;
6        *b = temp;
7    }
8
9    int main() {
10       int x = 50, y = 25;
11       std::cout << "Before swapping: x = " << x << ", y = " << y << std::endl;
12
13       swap(&x, &y);
14
15       std::cout << "After swapping: x = " << x << ", y = " << y << std::endl;
16
17       return 0;
18   }
19
```

Output:

```
"C:\Users\abhis\Google Drive\C++\pointerswap.exe"
Before swapping: x = 50, y = 25
After swapping: x = 25, y = 50

Process returned 0 (0x0)    execution time : 0.054 s
Press any key to continue.
```

Problems to Practice

1. Write a program that uses pointers to find sum of two integer variables. It adds the pointer values instead of the variable values.

2. Write a program with a function that takes in 3 numbers and arranges them in ascending order. It uses pointers to temporarily store values while swapping values.

3. Do the same as the above program to find the maximum value of an array of numbers.

Vectors

A vector is a collection of elements of the same data type. The number of elements in the vector is dynamic, meaning that it changes with every element that is added.

It is contained in the library class <vector> and can be invoked by

#include <vector>

A vector is defined as below:

vector<string> names;

The above line defines a vector of strings called names. When initialized, it contains 0 names. We can add names to the vector using push_back (which adds items to the back of the vector) or push_front(which adds items to the front of the vector)

Names.push_front("Abhi")

Adds Abhi to the front. So, the vector now has a length of 1.

Names.push_back("Bob")

Adds Bob to the back, which makes it the second element. Now, the vector has a length of 2. The 2 elements in order are "Abhi" and "Bob".

We can also remove items from the vector using pop_back or pop_front.

Names.pop_front() would remove "Abhi" while Names.pop_back() would remove "Bob".

Vector of Numbers

In the first vector program below, we create a vector of numbers and insert numbers into it in a while loop based on user input. The while loop checks if the user wants to continue adding numbers; and does so while the user says yes.

In line 5, the program declares a vector of integers called **numbers**.

In line 6, a character variable called **choice** is declared with a value of 'y'. The variable **choice** is used as a parameter for the while loop in line 8. The while loop runs as long as choice is equal to 'y'. In lines 11-12, the user inputs another value for **choice**, and the user inputs 'y' to continue adding numbers in the while loop.

In line 7, an integer variable called **newNumber** is created, and this variable holds the value of numbers that are input by the user. In line 10, the user inputs the value of **newNumber**. This value is added to the vector in line 11. The process in lines 9-11 occurs as long as the user wants to keep entering new numbers.

In lines 17-19, the programs output the values in vector **numbers** using a for loop. **numbers.size()** in line 17 is the length of the vector; and each element of the vector is accessed using **numbers[i]** in line 18.

```
1    #include <iostream>
2    #include <vector>
3
4    int main() {
5        std::vector<int> numbers;
6        char choice = 'y';
7        int newNumber;
8        while (choice == 'y') {
9            std::cout << "Enter a number: ";
10           std::cin >> newNumber;
11           numbers.push_back(newNumber);
12           std::cout << "Add another number? (y/n) ";
13           std::cin >>choice;
14       }
15
16       std::cout << "The numbers in the list are: ";
17       for (int i = 0; i < numbers.size(); i++) {
18           std::cout << numbers[i] << " ";
19       }
20       std::cout << std::endl;
21
22   }
23
```

Output:

In the sample output below, the users enters five numbers 5,10,45,78,34 and then hits a value of 'n' for choice to exit the while loop. The vector numbers are output as **[5 10 45 78 34]**.

In the example, we see that vectors are superior to arrays as the length of the array is constant and cannot adjust based on user preference. Vectors allow for dynamic allocation which is great for our example below where we don't know how many numbers are going to be entered by the user.

```
"C:\Users\abhis\Google Drive\C++\vectornumbers.exe"
Enter a number: 5
Add another number? (y/n) y
Enter a number: 10
Add another number? (y/n) y
Enter a number: 45
Add another number? (y/n) y
Enter a number: 78
Add another number? (y/n) y
Enter a number: 34
Add another number? (y/n) n
The numbers in the list are: 5 10 45 78 34

Process returned 0 (0x0)   execution time : 15.589 s
Press any key to continue.
```

Vector of Names

In below example, a vector of names is created. Two names are added, and one is removed. The vector is printed after each set of operations. This is a great example of how vectors are applied. If we used arrays for below example, we would have to change the size of the array each time a member was added or removed. In the case of vectors, it dynamically allocates the size.

In below code in line 5, a string vector called **vectornames** is created and has 5 names in it.

In line 7-11, the size of the vector and the list of names are printed.

In lines 14 and 15, two names, **Alice** and **Mike** are added to the vector.

In lines 16-20, the size of the vector and list of names are printed again. We see that the size of the vector increases to **7** and the two new names are added at the end (as **push_back** is used for both)

Now, one name is removed from the back of the vector using **pop_back** in line 23.

The size of the vector and list of names are again printed in lines 24-28. We see that the size is reduced to 6, and the last name Alice is removed from the vector.

All the outputs are shown in the output screen below the code.

```cpp
1   #include <iostream>
2   #include <vector>
3
4   int main() {
5       std::vector<std::string> vectornames = {"Alex", "Tim", "John", "Matthew", "Finny"};
6
7       std::cout << "Size: " << vectornames.size() << std::endl;
8       std::cout << "Contents: "<< std::endl;
9       for (int i = 0; i < vectornames.size(); i++) {
10          std::cout << vectornames[i] << " "<< std::endl;;
11      }
12      std::cout << std::endl;
13
14      vectornames.push_back("Mike");
15      vectornames.push_back("Alice");
16      std::cout << "Size 2: " << vectornames.size() << std::endl;
17      std::cout << "Contents2: "<< std::endl;
18      for (int i = 0; i < vectornames.size(); i++) {
19          std::cout << vectornames[i] << " "<< std::endl;;
20      }
21      std::cout << std::endl;
22
23      vectornames.pop_back();
24      std::cout << "Size 3: " << vectornames.size() << std::endl;
25      std::cout << "Contents 3: "<< std::endl;
26      for (int i = 0; i < vectornames.size(); i++) {
27          std::cout << vectornames[i] << " "<< std::endl;;
28      }
29
30      return 0;
31  }
32
```

154

Output:

```
■ "C:\Users\abhis\Google Drive\C++\vectorexample.exe"
Size: 5
Contents:
Alex
Tim
John
Matthew
Finny

Size 2: 7
Contents2:
Alex
Tim
John
Matthew
Finny
Mike
Alice

Size 3: 6
Contents 3:
Alex
Tim
John
Matthew
Finny
Mike

Process returned 0 (0x0)   execution time : 0.054 s
Press any key to continue.
```

Vector for student roll

In the next vector example, vectors are used to create a list of students and grades for a university. 5 vectors are created for names, math grades, science grades, social science grades and averages. The names, math grades, science grades, social science grades are input by the user in a while loop, and the average is calculated using the grades.

The vectors **names**, **mathGrades**, **scienceGrades**, **socialScienceGrades** and **averages** are created in lines 6-10. Names is a string vector and averages is a float vector, and the other three are integer vectors.

A character variable **choice** defined in line 12, is used to determine whether the while loop continues. It continues as long as the user enters a value of 'y' for **choice** in line 36.

In lines 20 to 30, the user inputs values for **newName**, **newMathGrade**, **newScienceGrade** and **newSocialScienceGrade**.

In line 32, **newAverage** is calculated using the three grades. All the new data is added to the vectors in line 31 and 33. This process continues till all the student data has been entered, and then user enters a value of 'n' for choice.

```
1    #include <iostream>
2    #include <string>
3    #include <vector>
4
5    int main() {
6        std::vector<std::string> names;
7        std::vector<int> mathGrades;
8        std::vector<int> scienceGrades;
9        std::vector<int> socialSciencesGrades;
10       std::vector<float> averages;
11
12       char choice = 'y';
13       while (choice == 'y') {
14           std::string newName;
15           int newMathGrade;
16           int newScienceGrade;
17           int newSocialSciencesGrade;
18           float newAverage;
19
20           std::cout << "Enter student name: ";
21           std::cin >>newName;
22           names.push_back(newName);
23           std::cout << "Enter math grade: ";
24           std::cin >>newMathGrade;
25           mathGrades.push_back(newMathGrade);
26           std::cout << "Enter science grade: ";
27           std::cin >>newScienceGrade;
28           scienceGrades.push_back(newScienceGrade);
29           std::cout << "Enter social sciences grade: ";
30           std::cin >>newSocialSciencesGrade;
31           socialSciencesGrades.push_back(newSocialSciencesGrade);
32           newAverage = (newMathGrade + newScienceGrade + newSocialSciencesGrade) / 3;
33           averages.push_back(newAverage);
34
35           std::cout << "Any more students? (y/n) ";
36           std::cin >> choice;
37       }
38
```

The number of students is now printed out in line 39 using **names.size()**.

The student's name and average are printed using the for loop in lines 42-44.

```
39       std::cout << "Number of students in class: " << names.size() << std::endl;
40       std::cout << "Student Roll: "<< std::endl;
41       for (int i = 0; i < names.size(); i++) {
42           std::cout << "Name:" << names[i] << " "<< std::endl;
43           std::cout << "Average:" << averages[i] << " "<< std::endl;
44       }
45
46       return 0;
47   }
48
```

Problems to Practice

1. Write a program that adds in bank customers data that is entered by the user. Have separate vectors for customer name, customer account balance, customer interest rate and customer credit rating.

2. Write a program that takes in data for gym memberships, with separate vectors for customer name, customer id number, customer joining date and membership expiration date. Calculate the days to expiration for each customer and put that in a separate vector.

3. Write a program that creates a vector for a grocery store. Each spot in the grocery store is an element on the vector, and it contains the name of the food in that grocery spot. The item can be removed a changed. If new spots are created to hold more items, new spots are added.

Maps

A map is a collection of a certain data type which is referenced by a key instead of an index.

A map is declared as shown below:

map<string, int> employeepay;

The line above declares a map called **employeepay**. Within the brackets, we see two variables. The first variable is a string and is the key to the map. The second variable is an integer variable and is the value in the map. In this case, the integer variable contains the salary of the employee. A map is assigned as shown below:

employeepay["Jim"]=52000;

In the above case, **Jim** is the string key to the **employeepay** integer value of **52000**.

A map is great when items are stored in an unordered fashion. In such cases, it's easier and more beneficial to access data using a map key instead of an index.

Map values can be accessed using square brackets [] or .at() function as shown below:

cout<<employeepay["Jim"]; or cout<<employeepay.at("Jim");

Both output a value of 52000.

The elements in a map are sorted by their keys. If the key is a string, they are sorted in ascending order.

The data is allocated in a dynamic fashion just like vectors, with the size changing each time an element is added or removed.

The main difference with vectors is that vectors are referenced by an index, while maps use a key.

Map Example 1

In our first example, we create two integer maps of weights and heights. Each map is accessed with a string key that represents the name of the person. The map values are assigned and printed out to show the basics of how a map works.

In code below, the two maps weight and height, are created in lines 5-6.

The maps are assigned integer data values in lines 8-9, and some of the values are printed out in lines 20-33.

```
1      #include <iostream>
2      #include <map>
3
4    ⊟int main() {
5          std::map<std::string, int> weight;
6          std::map<std::string, int> height;
7
8          weight["Steve"] = 85;
9          height["Steve"]=72;
10
11         weight["Abhi"] = 73;
12         height["Abhi"]=70;
13
14         weight["Bob"] = 93;
15         height["Bob"]=80;
16
17         weight["Zach"] = 65;
18         height["Zach"]=66;
19
20         std::cout << "Abhi's " << "weight is "<<weight.at("Abhi") << std::endl;
21         std::cout << "Steve's " << "height is "<<height.at("Steve") << std::endl;
22         std::cout << "Bob's " << "height is "<<weight["Steve"] << std::endl;
23
24
25         return 0;
26   }
27
```

Output:

The printed map values are shown below:

```
Abhi's weight is 73
Steve's height is 72
Bob's height is 85

Process returned 0 (0x0)     execution time : 0.043 s
Press any key to continue.
```

Printing out the Entire Map

To print the entire map, we use a for loop known as a range based for-loop. A range based for loop is used to access all elements in a map, as seen in lines 20-24.

In line 20, the **const auto &pair : weight** means that the variable **pair** will be used to access weight elements.

pair.first is the key to the map and **pair.second** is the value of the map that the key relates to.

So, **pair.first** is used to get the person's name and **pair.second** is the person's weight.

Since the second map height has the same key as the first map weight, we can use **pair.first** as the key. Using the key, we can access height using **height[pair.first]**. Using this, all key items are printed in lines 21-23 in the for loop. This is shown in the output below the code.

```cpp
1    #include <iostream>
2    #include <map>
3
4    int main() {
5        std::map<std::string, int> weight;
6        std::map<std::string, int> height;
7
8        weight["Steve"] = 85;
9        height["Steve"]=72;
10
11       weight["Abhi"] = 73;
12       height["Abhi"]=70;
13
14       weight["Bob"] = 93;
15       height["Bob"]=80;
16
17       weight["Zach"] = 65;
18       height["Zach"]=66;
19
20       for (const auto &pair : weight) {
21           std::cout << pair.first << "'s weight: " << pair.second << std::endl;
22           std::cout << pair.first << "'s height: " << height[pair.first] << std::endl;
23           std::cout << std::endl;
24       }
25
26
27       return 0;
28   }
29
```

Output:

One thing to note is that the data in the map is stored in the order of the keys. If the key is a string as in this example, the data is stored in alphabetical order. That's why Abhi's weight and height are printed first and Zach's are printed last.

```
Abhi's weight: 73
Abhi's height: 70

Bob's weight: 93
Bob's height: 80

Steve's weight: 85
Steve's height: 72

Zach's weight: 65
Zach's height: 66


Process returned 0 (0x0)   execution time : 0.050 s
Press any key to continue.
```

Removing Elements from a Map

Map data can easily be erased using the **erase()** function. This is seen in line 26 below when **Abhi's** weight is erased.

The new map data is printed out in lines 29-31.

As seen, even though the **height** data is not erased, it is not printed out for **Abhi** as it doesn't have access to the map key for pair defined in line 30. Pair.first has been erased for **Abhi,** so height[pair.first] does not print.

```cpp
#include <iostream>
#include <map>

int main() {
    std::map<std::string, int> weight;
    std::map<std::string, int> height;

    weight["Steve"] = 85;
    height["Steve"]=72;

    weight["Abhi"] = 73;
    height["Abhi"]=70;

    weight["Bob"] = 93;
    height["Bob"]=80;

    weight["Zach"] = 65;
    height["Zach"]=66;

    for (const auto &pair : weight) {
        std::cout << pair.first << "'s weight: " << pair.second << std::endl;
        std::cout << pair.first << "'s height: " << height[pair.first] << std::endl;
        std::cout << std::endl;
    }

    weight.erase("Abhi");

    std::cout <<"New Map Data:"<< std::endl;
    for (const auto &pair : weight) {
        std::cout << pair.first << "'s weight: " << pair.second << std::endl;
        std::cout << pair.first << "'s height: " << height[pair.first] << std::endl;
        std::cout << std::endl;
    }

    return 0;
}
```

Output:

```
Abhi's weight: 73
Abhi's height: 70

Bob's weight: 93
Bob's height: 80

Steve's weight: 85
Steve's height: 72

Zach's weight: 65
Zach's height: 66

New Map Data:
Bob's weight: 93
Bob's height: 80

Steve's weight: 85
Steve's height: 72

Zach's weight: 65
Zach's height: 66


Process returned 0 (0x0)   execution time : 0.047 s
Press any key to continue.
```

Searching for Elements in a Map

Elements in a map can be searched using the find function and assigned it to an iterator. An iterator in C++ is an object that acts as a pointer and allows the user to access elements in maps, lists etc.

```cpp
1    #include <iostream>
2    #include <map>
3
4    int main() {
5        std::map<std::string, int> weight;
6        std::map<std::string, int> height;
7
8        weight["Steve"] = 85;
9        height["Steve"]=72;
10
11       weight["Abhi"] = 73;
12       height["Abhi"]=70;
13
14       weight["Bob"] = 93;
15       height["Bob"]=80;
16
17       weight["Zach"] = 65;
18       height["Zach"]=66;
19
20       for (const auto &pair : weight) {
21           std::cout << pair.first << "'s weight: " << pair.second << std::endl;
22           std::cout << pair.first << "'s height: " << height[pair.first] << std::endl;
23           std::cout << std::endl;
24       }
25
26       weight.erase("Abhi");
27
28       std::cout <<"New Map Data:"<< std::endl;
29       for (const auto &pair : weight) {
30           std::cout << pair.first << "'s weight: " << pair.second << std::endl;
31           std::cout << pair.first << "'s height: " << height[pair.first] << std::endl;
32           std::cout << std::endl;
33       }
34
```

First, the program accepts a name from the user and stores it in the string variable **nametofind** in line 38 below.

The line **auto it = weight.find(nametofind)** creates an iterator that points to the name on the map that matches **nametofind**. If it cannot find the name, it returns **weight.end()**.

172

The if loop in line 41 checks if the name is in the map by checking if it is equal to **weight.end().** If it finds the name, it prints out the weight using it->second. it->second points to the weight of **nametofind**.

Since the height of the person has the same key/name, the program uses **height(nametofind)** to find the height. This is seen in line 42.

If it cannot find the name, it prints out **"Name not found"** in line 44.

```
35      std::cout << std::endl;
36      std::string nametofind;
37      std::cout<<"Enter name to search: ";
38      std::cin>>nametofind;
39      auto it = weight.find(nametofind);
40
41      if (it != weight.end()) {
42          std::cout << nametofind << "'s weight is " << it->second << " and height is " << height[nametofind]<<std::endl;
43      } else {
44          std::cout << nametofind << " not found." << std::endl;
45      }
46
47      return 0;
48  }
49
```

Output:

In the sample outputs below, the program first searches for Bob's name on user request. It finds the name and prints out his weight and height.

Then, the program is run again, and it searches for Abhi's name. The name cannot be found as it was deleted in the program in line 26.

```
Abhi's weight: 73
Abhi's height: 70

Bob's weight: 93
Bob's height: 80

Steve's weight: 85
Steve's height: 72

Zach's weight: 65
Zach's height: 66

New Map Data:
Bob's weight: 93
Bob's height: 80

Steve's weight: 85
Steve's height: 72

Zach's weight: 65
Zach's height: 66


Enter name to search: Bob
Bob's weight is 93 and height is 80

Process returned 0 (0x0)   execution time : 3.999 s
Press any key to continue.
```

```
Abhi's weight: 73
Abhi's height: 70

Bob's weight: 93
Bob's height: 80

Steve's weight: 85
Steve's height: 72

Zach's weight: 65
Zach's height: 66

New Map Data:
Bob's weight: 93
Bob's height: 80

Steve's weight: 85
Steve's height: 72

Zach's weight: 65
Zach's height: 66


Enter name to search: Abhi
Abhi not found.

Process returned 0 (0x0)   execution time : 3.345 s
Press any key to continue.
```

Problems to Practice

1. Write a program that keeps tracks of scores in a game of dice. It's a lot easier to use maps with keys as player names and just adding scores to each player.

2. Create a map that holds employee data and access employee data by employee id number. The maps contain employee names, salary, performance data.

3. Create a map that holds bank customer data with keys that can be accessed by bank customer number. The maps contain customer names, occupation, balance and credit rating.

Lists

A list is a C++ element that stores a series of elements. It is part of the standard library and can be accessed using **#include <list>.**

Lists are similar to arrays and maps. The main difference to arrays is that the length is dynamically allocated at runtime. So, the length of the list changes automatically at run time when elements are added or removed.

List elements are accessed using index elements. This differentiates them from maps that are accessed using keys.

A list has great utility when items have to processed in a particular order eg. customers in a queue, an online restaurant portal etc.

A list is defined as below:

list <string> names;

This creates a list of employees that contains string variables.

The variables are assigned using:

names.push_back("Sam") or names.push_front("Sam")

which pushes the item either to the back or front of the list respectively. Now, let's look at a few basic examples to understand this better.

List Example

In below example, list function is invoked from the C++ library in line 2.

In line 5, a list of strings called **queue1** is created.

In lines 6-8, three strings "Thomas", "Bob" and "Madison" are added to list queue1. In line 9, "Alvis" is added to the front of the queue.

The list is printed using the for loop in lines 11 to 13.

The for loop defined in line 11 goes through **queue1** list elements and assigns each iteration to element **i**. **i** is printed out in line 12 each iteration.

```cpp
1    #include <iostream>
2    #include <list>
3
4    int main() {
5        std::list<std::string> queue1;
6        queue1.push_back("Thomas");
7        queue1.push_back("Bob");
8        queue1.push_back("Madison");
9        queue1.push_front("Alvis");
10       std::cout << "Initial queue:"<<std::endl;
11       for (auto &i : queue1) {
12           std::cout << i << std::endl;
13       }
14       std::cout << std::endl;
15
16
17       return 0;
18   }
19
```

Output:

The output below has all the **queue1** list elements.

Madison is the last on the list since it is added using **push_back** and is the last element added.

Alvis is the last element added using **push_front** so it is the first on the list.

As shown below, lists are a great way to manipulate queues of data.

```
Initial queue:
Alvis
Thomas
Bob
Madison


Process returned 0 (0x0)   execution time : 0.045 s
Press any key to continue.
```

Adding and Removing Items from a List

In the next program, the first list example is modified by added 1 queue element and then removing one element. The list is printed after each change.

In line 16, a string called **Steve** is added to queue1 list to the 2nd spot. The **++queue.begin()** implies that Steve is added to beginning of the queue + 1 person, or 2nd person in the queue. The queue is then printed in lines 18-20.

In line 24, a person is removed from the front of the list using **"pop_front"**. This removes **Alvis** as he is the first list member.

```cpp
1    #include <iostream>
2    #include <list>
3
4    int main() {
5        std::list<std::string> queue1;
6        queue1.push_back("Thomas");
7        queue1.push_back("Bob");
8        queue1.push_back("Madison");
9        queue1.push_front("Alvis");
10       std::cout << "Initial queue:"<<std::endl;
11       for (auto &i : queue1) {
12           std::cout << i << std::endl;
13       }
14       std::cout << std::endl;
15
16       queue1.insert(++queue1.begin(), "Steve");
17       std::cout << "Queue after 1 person pushes Bob to get 2nd spot "<<std::endl;
18       for (auto &i : queue1) {
19           std::cout << i << std::endl;
20       }
21       std::cout << std::endl;
22
23
24       queue1.pop_front();
25       std::cout << "Queue after 1st person served"<<std::endl;
26       for (auto &i : queue1) {
27           std::cout << i << std::endl;
28       }
29       std::cout << std::endl;
30
31
32       return 0;
33   }
34
```

184

Output:

In the output below, we see **Steve** added as the second person the list
to displace **Bob**.

Then, **Alvis** is removed when **pop_front** is used to remove the person
in front of the list.

```
■ "C:\Users\abhis\Google Drive\C++\list.exe"

Initial queue:
Alvis
Thomas
Bob
Madison

Queue after 1 person pushes Bob to get 2nd spot
Alvis
Steve
Thomas
Bob
Madison

Queue after 1st person served
Steve
Thomas
Bob
Madison


Process returned 0 (0x0)    execution time : 0.059 s
Press any key to continue.
```

185

Auto Dealership List

In the next example, a list is used to service customers in an auto dealership. The person in the front is serviced when it's his turn and he's removed from the list using **pop_front()**. New customers are added to the back of the list using **push_back()**.

The list **customerQueue** is created as a string list in line 4.

3 customer names are added as strings to **customerQueue** in lines 6-8, and this list is printed in lines 10-13.

One customer is serviced in line 15. This removes a person from the front of the list. The list is then printed in lines 17-20. In the output below the code, we see that **John Smith** is removed from the list.

Now, the program accepts a string called **newCustomer** from the user. **newCustomer** is added to the back of the **customerQueue**, and the list is again printed in lines 30-32.

```cpp
#include <iostream>
#include <list>
int main() {
    std::list<std::string> customerQueue;

    customerQueue.push_back("John Smith");
    customerQueue.push_back("Jane Doe");
    customerQueue.push_back("Bob Johnson");

    std::cout << "Current queue:" << std::endl;
    for (const auto& customer : customerQueue) {
        std::cout << customer << std::endl;
    }

    customerQueue.pop_front();
    std::cout << std::endl;
    std::cout << "Updated queue after serving 1st customer:" << std::endl;
    for (const auto& customer : customerQueue) {
        std::cout << customer << std::endl;
    }

    std::cout << std::endl;
    std::string newCustomer;
    std::cout << "Enter the name of the new customer: ";
    std::cin >>newCustomer;
    customerQueue.push_back(newCustomer);

    std::cout << std::endl;
    std::cout << "Updated queue:" << std::endl;
    for (const auto& customer : customerQueue) {
        std::cout << customer << std::endl;
    }
    return 0;
}
```

Output:

In the output below, we add Joe as the new customer, and Joe is added to the back of **customerQueue**.

```
"C:\Users\abhis\Google Drive\C++\autodealerqueue.exe"

Current queue:
John Smith
Jane Doe
Bob Johnson

Updated queue after serving 1st customer:
Jane Doe
Bob Johnson

Enter the name of the new customer: Joe

Updated queue:
Jane Doe
Bob Johnson
Joe

Process returned 0 (0x0)   execution time : 7.073 s
Press any key to continue.
```

Problems to Practice

1. Write a program that represents a list of customers for an airport queue. It includes a list for customer name, customer check-in time and check out time. It can remove customers and add new customers; and it keeps track of total number of customers in the lounge at the same time.

2. Write a program that deals with customer complaints for an online business. There's a queue of complaints. Create two lists: one for customer names, another for customer products and the final one for refund amount. All of these are changed dynamically as customer complaints are dealt with.

What is an object?

An object is a data structure that contains data and functions that operate on the data. The data may be of the same of the same or different variable type. These objects are great to model real world applications.

For example, we may want to store car data, which consists of make, model, mileage, tyre diameter, engine rpm etc. And we may want to have functions that use the data to calculate other variables like reliability, weight etc.

What are classes?

Classes are the best blueprints for creating objects. Classes have a structure for creating a variety of data types and functions that utilize the data.

Classes can have both a public and private component. Private variables and functions cannot be accessed outside the class, while public variables and functions can be accessed outside the class using an object. Now, let's look at a simple example to make all a lot clearer.

A Basic Class

In this program, we create a class called Circle that has two public functions and one private variable. This private variable is an input when an instance of the class is created in the main program.

The class is first initialized in lines 4-6 with one input variable called **radius**. This is a private variable, meaning that it can only be accessed within the class. **getArea()** and **getCircumference()** are public functions within the class.

The private variable **radius** is created in line 15-17.

getArea() in lines 7-10 calculates the area of the circle with radius value of float variable **radius**.

getCircumference() in lines 11-14 calculates the circumference of the circle with radius value of float variable **radius**.

In the main program starting line 20, the user enters the value of radius, and it is stored in float variable **r**.

In line 24, an instance of the class called **myCircle** is created with an input radius of **r**.

MyCircle cannot access any private variable of the class, but it can access the public functions using **MyCircle.getArea()** and **MyCircle.getCircumference().**

The area obtained by **MyCircle.getArea()** is stored in a float variable called **area** in line 25, and it is printed on the screen in line 27.

The circumference obtained by **MyCircle.getCircumference()** is stored in a float variable called **crcm** in line 26, and it is printed on the screen in line 28.

```
1    #include <iostream>
2    class Circle {
3    public:
4        Circle(float radius):radius(radius)
5        {
6        }
7        float getArea()
8        {
9            return 3.14 * radius * radius;
10       }
11       float getCircumference()
12       {
13           return 2 * 3.14 * radius;
14       }
15   private:
16       float radius;
17   };
18
19   //now using this class
20   int main() {
21       float r;
22       std::cout << "Enter radius";
23       std::cin >> r;
24       Circle myCircle(r);
25       float area = myCircle.getArea();
26       float crcm = myCircle.getCircumference();
27       std::cout << "The area of the circle is: " << area << std::endl;
28       std::cout << "The circumference of the circle is: " << crcm << std::endl;
29       return 0;
30   }
31
```

Output:

All the data is seen in the output below the code. The user enters a radius value of 7. It is stored in float variable r.

The value of the area is calculated with this radius vale using **myCircle.getArea()** and output printed below.

The value of the area is calculated with this radius vale using **MyCircle.getCircumference()** and output printed below.

```
■ "C:\Users\abhis\Google Drive\C++\Rectangle.exe"

Enter radius 7
The area of the circle is: 153.86
The circumference of the circle is: 43.96

Process returned 0 (0x0)    execution time : 3.992 s
Press any key to continue.
```

Basic Class Example 2

In the next example, we create a class called Dog which has 4 public variables and 4 public functions. The class basically takes in the dog name, breed, location and age and prints them out using public member functions.

In the code below, the class is first initialized in line 8-9 as a class with four input variables.

The four public variables are declared in lines 4-7. They consist of 3 string variables **name**, **breed** and **location** and one integer variable **age**.

The four public functions are declared in lines 11-27. The functions are **getname(), getbreed(),getage()** and **getlocation().** They return the values of all the four variables mentioned.

The main program starts in line 31. An instance of the class is created in line 32. The instance is called **Dog1** and enters values for all 4 variables. **Dog1** has a name of "Cleo", a breed of "Dalmation", an age of 11, and a city of "Atlanta".

The four member functions of the class are accessed using **Dog1.getname(), Dog1.getbreed, Dog1.getage() and**

Dog1.getlocation(). They are stored in string variables **dn**, **da**, **db** and

dl in lines 33-36.

These values are printed out in lines 37-40.

This is seen in the output below the code.

```
1    #include <iostream>
2    class Dog {
3        public:
4            std::string name;
5            std::string breed;
6            int age;
7            std::string location;
8            Dog(std::string name, std::string breed, int age, std::string location) : name(name), breed(breed), age(age), location(location)
9            {
10           }
11           std::string getname()
12           {
13               return name;
14           }
15           std::string getbreed()
16           {
17               return breed;
18           }
19           int getage()
20           {
21               return age;
22           }
23           std::string getlocation()
24           {
25               return location;
26           }
27   };
28
29
30   //now using this class
31   int main() {
32       Dog dog1("Cleo", "Dalmation", 11, "Atlanta");
33       std::string dn=dog1.getname();
34       std::string db=dog1.getbreed();
35       int da=dog1.getage();
36       std::string dl=dog1.getlocation();
37       std::cout << "The Dog's name is " << dn << std::endl;
38       std::cout << "The Dog's age is " << da << std::endl;
39       std::cout << "The Dog's breed is " << db << std::endl;
40       std::cout << "The Dog's location is " << dl << std::endl;
41
42   }
43
```

Output:

std::string name;

"C:\Users\abhis\Google Drive\C++\Dog_Class.exe"

```
The Dog's name is Cleo
The Dog's age is 11
The Dog's breed is Dalmation
The Dog's location is Atlanta

Process returned 0 (0x0)   execution time : 0.046 s
Press any key to continue.
```

Create a Bank Account

In the next example, we create a class called BankAccount that imitates the functions of a bank. The class has 3 public variables and 4 public member functions. It takes in customer name, bank balance and account number as public variables. The 4 class member functions are used to deposit money, withdraw money, print account balance and print account number.

In the program below, the class is initialized in line 8 as a class with 4 variables.

The three variables are declared in lines 4-6. There is one string variable **holdername** that holds the customer's name and two variables **balance** and **accountNum**. These hold the funds balance and account number respectively.

The deposit function in lines 11-14 accepts a float variable **amount** as input and adds that to the float variable **balance**. It basically performs the function of a customer deposit.

The withdraw function in lines 16-19 accepts a float variable **amount** as input and subtracts that from the float variable **balance**. It basically performs the function of a customer withdrawal.

The other two functions **getBalance()** and **getAccountNum()** basically output the **balance** and **accountNum** variables.

The main function starts in line 36. In line 37, an instance of the class called **BankAccount1** is created with a **balance** of 5000, a **holdername** of "John Smith" and **accountNum** of 007123.

These variables are printed out in lines 38-39 using the **getBalance()** and **getAccountNum()** functions in the class. The public variable **holderName()** is also used.

In line 41, a deposit of 500 is made; and the new balance is printed in line 42.

Two withdrawals are made in line 44 and 50. If statements are used to check if the withdrawal is more than the balance. If the withdrawal is more than the balance, an error message is printed. If not, the transaction is completed.

The final customer balance is printed in each case.

All the screen outputs are shown below the code, and you can run it yourself to verify.

```cpp
#include <iostream>
class BankAccount {
    public:
        float balance;
        int accountNum;
        std::string holderName;

        BankAccount(double balance, int accountNum, std::string holderName): balance(balance), accountNum(accountNum), holderName(holderName) {}

        void deposit(float amount)
        {
            balance = balance + amount;
        }

        bool withdraw(float amount)
        {
            if (amount > balance) {
                return false;
            }
            balance = balance - amount;
            return true;
        }

        double getBalance() {
            return balance;
        }

        int getAccountNum() {
            return accountNum;
        }
};


int main() {
    BankAccount account1(5000, 007123, "John Smith");
    std::cout << "Account Holder: " << account1.holderName << std::endl;
    std::cout << "Initial balance for Account " << account1.getAccountNum() << ": " << account1.getBalance() << std::endl;

    account1.deposit(500);
    std::cout << "Balance after deposit: " << account1.getBalance() << std::endl;

    if (account1.withdraw(2000)) {
        std::cout << "Withdrawal successful. New balance: " << account1.getBalance() << std::endl;
    } else {
        std::cout << "Insufficient funds. Current balance: " << account1.getBalance() << std::endl;
    }

    if (account1.withdraw(1000)) {
        std::cout << "Withdrawal successful. New balance: " << account1.getBalance() << std::endl;
    } else {
        std::cout << "Insufficient funds. Current balance: " << account1.getBalance() << std::endl;
    }

    return 0;
}
```

201

Output:

In the output below, the program first prints out the account holder name John Smith and an account balance of 5000. After adding 500, the account balance printed out is now 5500.

After the first withdrawal of 2000, the account balance is now 3500.

After the second withdrawal of 1000, the account balance updates to 2500.

```
"C:\Users\abhis\Google Drive\C++\bankaccount.exe"
Account Holder: John Smith
Initial balance for Account 3667: 5000
Balance after deposit: 5500
Withdrawal successful. New balance: 3500
Withdrawal successful. New balance: 2500

Process returned 0 (0x0)    execution time : 0.057 s
Press any key to continue.
```

Problems to Practice

1. Write a program that adds a function to above bank program. It adds a reward of 10% to the balance to anyone who wishes to deposit more than 10000.

2. Create a class called Car that takes in car name, car brand, number of miles, weight and cost. Create member functions that can output all above data.

3. Create a class Rectangle that takes in a width and height and can return area and perimeter.

4. Create a class that simulates a bank loan. It takes in credit rating as either A, B, C, D. It takes in a name, a loan amount and an account number. It has 2 functions: one which calculates the annual interest rate; and one which output both interest rate and annual interest added.

   a. Credit rating of 'A' has interest rate of 2%; 'B' has 4% rate; C has 6% rate; D has an 8% rate

   b. Interest is calculate using simple interest formula of:

   Interest = (Loan Amount * Time * Interest Rate) / 100 ;

   Time is 1 year for annual interest rate.

204

Common C++ Code Errors

While running C++, it's normal to encounter errors in the code that need fixes. It's part of the fun of coding. Programmers of all levels encounter errors. However, there are a few basic errors that are most common in C++. We've listed out 5 errors that you are most likely to encounter in C++ as a beginner.

1. Syntax Error

This error refers to an error in the syntax of your code, such as a missing semicolon (;) after a statement.

The example below would have a syntax error due to the missing semicolon in line 4.

```
1    #include <iostream>
2
3    int main() {
4        std::cout << "Hello, World!" << std::endl
5        return 0;
6    }
7
```

2. Undeclared Identifier

This error occurs when a variable is used by a program that hasn't been defined yet.

In example below, variable **name** has not been defined, but the code tries to print its value in line 6. This would cause an undeclared identifier error.

```
1    #include <iostream>
2
3    int main() {
4
5        std::cout << "Hello, World!" << std::endl;
6        std::cout << "My name is " << name << std::endl;
7        return 0;
8    }
9
```

To prevent this, we change the code as follows, and declare a string **name** in line 6.

```
1    #include <iostream>
2
3    int main() {
4
5        std::cout << "Hello, World!" << std::endl;
6        std::string name="Dave";
7        std::cout << "My name is " << name << std::endl;
8        return 0;
9    }
10
```

3. Undefined reference

An undefined reference occurs when an item is referenced in the code that has not been defined.

In the below example, **printname()** has not been defined properly in line 3, but it is referenced in line 7.

```
1     #include <iostream>
2
3     void printname();
4     int main() {
5         std::cout << "Hello, World!" << std::endl;
6         std::string name="Dave";
7         printname();
8         return 0;
9     }
10
```

We correct this by properly defining **printname()** below. **Printname** is a function that accepts a string as a parameter and prints it out as seen lin lines 4-6.

```
1    #include <iostream>
2
3    void printname(std::string name)
4    {
5         std::cout << name << std::endl;
6    }
7    int main() {
8         std::cout << "Hello, World!" << std::endl;
9         std::string name="Dave";
10        printname(name);
11        return 0;
12   }
13
```

4. Uninitialized variable

An uninitialized variable error occurs when a variable is declared and referenced in the code but no value has been assigned to it yet. In the example below, integer variable **tst** has not been initialized in line 5 but is used in line 6.

```
1    #include <iostream>
2
3    int main()
4    {
5         int tst;
6         std::cout << "The value of tst is: " << tst << std::endl;
7         return 0;
8    }
9
```

To correct this, we add a value to **tst** in line 5.

```
1    #include <iostream>
2
3    int main()
4    {
5        int tst=5;
6        std::cout << "The value of tst is: " << tst << std::endl;
7        return 0;
8    }
```

5. Type Mismatch

A type mismatch error occurs when two variables of incompatible variable types are combined. For example, a string variable is set equal to an integer variable or vice versa.

In the code below, there is one integer variable x, and one string variable y. x is set equal to y in line 7. This returns a type mismatch error as the two variables are incompatible.

```
1    #include <iostream>
2
3    int main()
4    {
5        int x = 10;
6        std::string y = "Steve";
7        x = y;
8        return 0;
9    }
```

US States and Capitals array

```
std::string states[] =
{"Alabama","Alaska","Arizona","Arkansas","California","Colorado","Co
nnecticut","Delaware","Florida","Georgia","Hawaii","Idaho","Illinois","I
ndiana","Iowa","Kansas","Kentucky","Louisiana","Maine","Maryland","
Massachusetts","Michigan","Minnesota","Mississippi","Missouri","Mo
ntana","Nebraska","Nevada","New Hampshire","New Jersey","New
Mexico","New York","North Carolina","North
Dakota","Ohio","Oklahoma","Oregon","Pennsylvania","Rhode
Island","South Carolina","South
Dakota","Tennessee","Texas","Utah","Vermont","Virginia","Washingto
n","West Virginia","Wisconsin","Wyoming"};
```

```
std::string capitals[] = {"Montgomery","Juneau","Phoenix","Little
Rock","Sacramento","Denver","Hartford","Dover","Tallahassee","Atlan
ta","Honolulu","Boise","Springfield","Indianapolis","Des
Moines","Topeka","Frankfort","Baton
Rouge","Augusta","Annapolis","Boston","Lansing","St.
Paul","Jackson","Jefferson City","Helena","Lincoln","Carson
City","Concord","Trenton","Santa
Fe","Albany","Raleigh","Bismarck","Columbus","Oklahoma
City","Salem","Harrisburg","Providence","Columbia","Pierre","Nashvill
e","Austin","Salt Lake
City","Montpelier","Richmond","Olympia","Charleston","Madison","Ch
eyenne"};
```

The end... almost!

Reviews are not easy to come by.

As an independent author with a tiny marketing budget, I rely on readers, like you, to leave a short review on Amazon.

Even if it's just a sentence or two!

So if you enjoyed the book, please click this link and leave a review...

I am very appreciative for your review as it truly makes a difference.

Thank you from the bottom of my heart for purchasing this book and reading it.

www.ingramcontent.com/pod-product-compliance
Lightning Source LLC
Chambersburg PA
CBHW061020220326
41597CB00016BB/1715